Positive Computing

Positive Computing

Technology for Wellbeing and Human Potential

Rafael A. Calvo and Dorian Peters

The MIT Press
Cambridge, Massachusetts
London, England

© 2014 Massachusetts Institute of Technology

All rights reserved. No part of this book may be reproduced in any form by any electronic or mechanical means (including photocopying, recording, or information storage and retrieval) without permission in writing from the publisher.

MIT Press books may be purchased at special quantity discounts for business or sales promotional use. For information, please email special_sales@mitpress.mit.edu.

This book was set in ITC Stone Serif Std 9/13pt by Toppan Best-set Premedia Limited. Printed and bound in the United States of America.

Library of Congress Cataloging-in-Publication Data

Calvo, Rafael A.
Positive computing : technology for wellbeing and human potential / Rafael A. Calvo and Dorian Peters.
 p. cm.
Includes bibliographical references and index.
ISBN 978-0-262-02815-8 (hardcover : alk. paper)
1. Computers—Social aspects. 2. Human-computer interaction. 3. Well-being. 4. Positive psychology. 5. Neurosciences. 6. Technology—Social aspects. I. Peters, Dorian, 1977– II. Title.
QA76.9.C66C354 2014
303.48'34—dc23
2014010627

10 9 8 7 6 5 4 3 2 1

For our parents and children. May the technology in our future help us all to be well, happy, and wise.

Contents

Acknowledgments   ix

1 An Introduction to Positive Computing   1

I   11

2 The Psychology of Wellbeing   13
3 Multidisciplinary Foundations   41
4 Wellbeing in Technology Research   63
5 A Framework and Methods for Positive Computing   81

II   107

6 Positive Emotions   109
7 Motivation, Engagement, and Flow   131
8 Self-Awareness and Self-Compassion   155
9 Mindfulness   179
10 Empathy   203
11 Compassion and Altruism   229
12 Caveats, Considerations, and the Way Ahead   257

Index   281

Acknowledgments

We are grateful to many people and organizations: to Robert Dale, Alan Blackwell, Yvonne Rogers, Peter Goodyear, Douglas Sery, and, of course, the proposal and book reviewers who provided us with invaluable feedback at different stages; to the experts who generously provided their perspectives as sidebars throughout the book; to family for babysitting and seeds planted long ago; to Francisco and Eva for their inspiration.

Rafael was kindly hosted by Peter Robinson and the Computer Laboratory at the University of Cambridge for much of the writing of this book. Rafael also acknowledges the Young and Well Cooperative Research Centre and Inspire Foundation, from which we constantly learn about the challenges faced by organizations using technology to improve wellbeing and the world around us.

Sidebars

We gratefully acknowledge the contribution of the following experts whose perspectives are included in sidebars throughout the chapters of this book.

Jeremy Bailenson, Stanford University. "Virtual Altruism"—chapter 11
Timothy W. Bickmore, Northeastern University. "Compassionate Computer Agents"—chapter 10
danah boyd, Harvard and Microsoft. "Making Sense of Increased Visibility"—chapter 3
Jane Burns, Young and Well Cooperative Research Centre. "When Worlds Collide: The Power of Cooperation in Wellbeing Science"—chapter 3
David R. Caruso, Yale University and EI Skills Group. "How Emotional Intelligence Can Inform Positive Computing"—chapter 8

Mihaly Csikszentmihalyi, Claremont Graduate University. "Computers and the Condition of Human Plurality"—chapter 10

Felicia Huppert, University of Cambridge. "Measuring Subjective Wellbeing"—chapter 2

Mary-Helen Immordino-Yang, University of Southern California. "Developing Computer Interfaces That Inspire: Insights from Affective Neuroscience"—chapter 11

Adele Krusche and J. Mark G. Williams, University of Oxford. "Mindfulness Online"—chapter 9

Jane McGonigal, Institute for the Future. "Let the Positive Games Begin"—chapter 6

Jonathan Nicholas, Inspire Foundation. "Inspiring Projects—Opportunities for Mental Health and Technology"—chapter 2

Don Norman, Nielsen Norman Group. "Fun and Pleasure in Computing Systems"—chapter 6

Yvonne Rogers, University College London. "Is a Diet of Data Healthy?"—chapter 4

1 An Introduction to Positive Computing

"Don't be evil," wrote Larry Page and Sergey Brin on the eve of Google's IPO in 2004. Almost a decade later, Apple CEO Tim Cook opened the annual developer's conference with a tribute to emotional experience as part of a campaign in which Apple claimed to ask of their technologies: "Will it make life better? Does it deserve to exist?"

These messages, however aspirational, resonate as overarching goals for a growing number of technologists who want to ensure the work they love to do is actively improving people's lives. If a technology doesn't improve the wellbeing of individuals, society, or the planet, should it exist?

The desire to "do good with technology" has emerged from a shared experience that technology has a major impact on how we live, that it has the capacity not only to increase stress and suffering, but also to improve lives individually and en masse. Indeed, the potential influence of digital and ubiquitous technologies is unprecedented. As you read this book, there are more *mobile* devices than people on the planet,[1] and over the past decade we have watched them play a starring role in the politics of nations, in the politics of human relationships, and in the day-to-day social and emotional dynamics of our lives.

As a result, a growing number of technology professionals are seeking a realignment of business goals away from profit and toward social good—a sentiment manifest in the advent of social enterprise that places profit making secondary to a social purpose.[2] Within the technology industry, we have seen the emergence of initiatives such as Games for Change, UX for Good, Wisdom 2.0, and Design for Good, while human–computer interaction (HCI) conferences provide ongoing testament to the growth in HCI for wellbeing, social impact, and peace.

This growing interest in social good among technology professionals is part of a larger emerging public concern for how our digital experience is impacting our emotions, our quality of life, and our happiness. We are gradually leaving behind the stark mechanical push for productivity and efficiency that characterized the early age of computing and maturing into a new era in which people demand that technology contribute to their wellbeing as well as to some kind of net social gain.

This sentiment reflects a broader renaissance of focus on humanistic values such as happiness and human potential that has begun to flourish across many different disciplines. A shift in priorities is now loud and clear among economists, politicians, and policymakers as they turn to statistical measures of wellbeing and "gross national happiness" as new indicators of success (Helliwell, Layard, & Sachs, 2012).[3]

Similarly, in the past decade psychologists and psychiatrists have achieved hard-won disciplinary support for research that goes beyond illness into aspects of healthy functioning such as resilience, happiness, and altruism.[4] In concert, neuroscientists have been exploring the physiology of exceptionally healthy minds and studying constructs such as empathy, mindfulness, and meditation empirically. Their findings are fueling action by educators and business leaders who are applying work on emotional intelligence and positive psychology to improve wellbeing among their students and workers (Joinson, McKenna, Postmes, & Reips, 2007; Ong & van Dulmen, 2006). It's inevitable that technology should begin to play a more sophisticated part in these multidisciplinary efforts toward supporting wellbeing.

In this book, we refer to this area of work—the design and development of technology to support psychological wellbeing and human potential—as "positive computing."[5] We believe we are seeing the beginning of an important shift in the focus of modern technologies in which multidisciplinary efforts to support human flourishing are helping to shape thinking around how we design for digital experience.

In the same way that economists are measuring wellbeing at the national level and psychologists have been measuring it at an individual level for decades, it's time to consciously and systematically consider wellbeing measures in the design and evaluation of technology.

That isn't to say it will be easy. Understanding the impact of technology on individuals and on society is fraught with the challenges common to understanding any highly complex system. Cultural, social, ethical, and psychological variables will inevitably conspire to create a complex, nuanced, and challenging space for investigation. This suggests that partnering with social scientists (old hands at dealing empirically with multifaceted human systems) will be absolutely vital to success.

A simple glance at modern media suggests that the public is eager for those of us working in technology to take on this challenge. Best-seller lists abound with books on happiness as well as with books on how technology is affecting it. Warnings that technology is degrading our intelligence and inducing stress sit alongside promises of how it will save the world.[6]

What's clear is that many of us are interested in (and even nervous about) how these pervasive tools are affecting us, and we seek out ways of getting a handle on the situation. After all, it's arguably our fundamental goal in life as human beings to pursue happiness, and in the modern world we're either going to do so with the help of technology or in spite of it.

Of course, some might argue that technological progress in and of itself is enough to improve wellbeing across the population. Tempting as it is to go along with that assumption, the evidence persists in suggesting otherwise.

Technological Progress a Poor Proxy for Wellbeing

Remarkably, despite major advancements and an incredible proliferation of devices, there is no evidence our modern tools have made us psychologically healthier or happier today than we were 20 years ago.[7] In other words, just as wealth has proven to be an inadequate proxy for a nation's wellbeing (Kahneman, Krueger, Schkade, Schwarz, & Stone, 2006), technology use has proven similarly inadequate as a surrogate for increased wellbeing in an individual or across a population.

However, if digital technologies are not actively supporting our wellbeing, it is simply because we have yet to consider it in the design cycle of technology. This oversight has occurred for many reasons, including a historical position among engineers and computer scientists that makes us more comfortable staying clear of the difficult-to-quantify and value-laden aspects of psychological impact. In other words, wellbeing has been not only traditionally overlooked but even consciously excluded from consideration owing to a legacy of industry discomfort with certain aspects of humanness.

The Human-Machine Legacy

Although there are significant exceptions (including the critical work in values-sensitive design and other efforts in HCI), few technology professionals have begun adding user wellbeing to their brief. In fact, just try mentioning happiness or wellbeing at a seminar for software engineers, and you're guaranteed some eyebrow raising (or that special breed of academic heckling). In his book *Texture*, Richard H. R. Harper (2010) attributes the pragmatic and behaviorist mindset that dominates HCI to the early influence of Alan Turing and Norbert Wiener.

> *Turing believed he was inventing a new discipline, one that dealt with algorithms. But this vision also included a view of the human. As it happens, Wiener thought that the science he was inventing, cybernetics, was all about people, even though his science was enormously mathematical, and hence quite close to what Turing thought he was doing. But the world view that these individuals have produced is one in which people—the users—turn out to be not very human at all. They have humanlike capacities and humanlike behaviors to be sure, but they are so reduced in their sensibilities that the humanness has been taken out. … Those who adopt Turing's view assume that what goes on inside the [human] machine itself is not only invisible but also somehow tricky and best avoided.*

Harper makes no claim to being immune to this influence, and he shares insightful examples of technology development projects that failed to predict actual human use: "We recognized that issues of human action were relevant here, but our instincts were to avoid them; Turing's aversion to moral overtones encouraged us away." He goes on to add that the public has likewise learned to think in the same way about themselves: "people think of themselves as machines … and worry about optimizing their performance."

In attempting to unpack the Turing–Wiener legacy further, we have found that researchers in the previously mentioned eyebrow-raising category generally take one of four positions:

• The first (which we call the "positive dogmatic") argues that technological progress itself is enough to make the world a better place; as countries become more technologically advanced, they become wealthier, more educated, and healthier, and when you're rich, educated, and healthy, you'll be happy. In this view, a technological singularity leads to utopia.
• A second group (which we call the "negative dogmatic" or "modern luddite") believes that technology is inherently negative because it increases things such as unemployment, loneliness, stress, depression, and so on.
• The third group (agnostics) argue that it is impossible to say if products have positive or negative effects on wellbeing because, as a matter of principle, you can't measure that kind of thing.
• And finally, the fourth group is what Sextus Empiricus would call "skeptics" because they question the very existence of concepts such as wellbeing, happiness, and emotion in the first place.

Although we believe this is changing, we have found that our industry's traditional view of humans, useful as it has been to invention historically, creates barriers to progress as we get to a point where devices—far from being the mammoth expert-handled machines they once were—have become embedded into the daily experiences that *shape* all of us.

Nevertheless, some technologists remain reluctant to go beyond the apparent safety of a machine view of users. They are understandably wary of the empirical challenges that this change presents and skeptical of the feasibility of delving into psychological and subjective issues such as wellbeing within their field—and rightly so, as it is entirely true that the technology field alone is not equipped for such a task. It has neither sufficient experience nor appropriate methodologies for dealing with the complexities of human psychological wellbeing, which is why *multidisciplinary partnership is crucial*. Partnerships with psychologists, anthropologists, sociologists, and educational researchers are already common within branches of HCI, so following in these footsteps should not require too drastic a leap.

Overall, in discussions with field leaders about positive computing, we find our colleagues are most likely to feel uncertain about the feasibility of *measuring* a concept as apparently nebulous and personal as wellbeing. Fortunately, on that point social science has spent decades refining instruments for precisely this purpose.

Measuring What Matters

Indeed, positive computing may appear out of reach at first glance, in the way that "user experience" felt fuzzy and impractical at the turn of the millennium. Although in technology fields we have little experience with measuring psychological impact, fields such as psychology and psychiatry proffer a wealth of empirically validated methodologies and best practice prerequisite to taking on this challenge.

For example, researchers have been measuring and assessing attributes such as happiness, quality of life, and subjective wellbeing since at least the 1970s (Fordyce, 1977). There are now more than 1,400 wellbeing and quality-of-life instruments for various specific subgroups (customized to age, culture, religion, context, etc.) and thousands of studies validating these instruments.[8]

Two of the most widely used measures of wellbeing are the Center for Epidemiological Studies–Depression (CES-D) Scale, which has been used in more than 23,000 studies, and the Global Assessment of Functioning Scale used by psychiatrists and psychologists in clinical and research settings. Doctors, insurance companies, and government agencies rely on these measures to make decisions about treatment, benefits, and spending.

Recent technological advances in areas such as affective computing, computer vision, and data mining are also making inroads. Technology can now help us to better understand people's emotional experience through the analysis of text, facial expression, physiology, interaction, and behavioral analytics. We can also learn from research in cyber-therapies and educational technologies, both of which seek to combine information about user behavior, cognition, and affect to inform their work.

Research and practice in medicine and the social sciences have shown us that measuring wellbeing and related factors not only is entirely feasible but has been well established for a number of decades. But is there any evidence that the technologies we build might actually be recruited to have a positive impact on wellbeing? Again, the work of psychologists has paved the way.

Studies in psychology have already combined the use of wellbeing measures with digital technologies for the delivery of Internet-based "interventions" (interventions are therapeutic or promotional efforts to improve mental or physical health). The *Journal of Medical Internet Research* and the *Journal of Cyberpsychology, Behavior, and Social Networking* are two of the most highly ranked journals publishing in this area. *IEEE Transactions on Affective Computing* also publishes research on the emotional impact of computers, but from an engineering perspective. Psychology research continues to uncover strategies empirically shown to lead to increases in long-term wellbeing, many of which are detailed later in this book.

Although psychologists have developed many proven ways to strengthen our mental resources, we spend much more time with digital technologies than we do with psychologists; digital technologies have unparalleled demographic reach. As psychology researchers Stephen Schueller and Acacia Parks (2012) have said, "The science of internet [sic] interventions can be advanced through expanding options and strategies to promote worldwide wellbeing."

As an example of sheer numbers, in 2012 researchers at Facebook published a study in *Nature* that measured the impact of three interface-design variations on social participation behavior (Bond et al., 2012). This randomized control trial had a whopping 61 million participants and succeeded in showing how a small design change can have impressive consequences on user thinking and behavior.

We currently go about designing new technologies without any sense of how our design decisions will impact our users' psychological health and flourishing. Imagine the effects of taking that aspect into account, even just a little bit. Wellbeing-driven improvements to digital experiences have the unique potential to effect population-wide positive change.

Developments in the field of positive computing will have the side effect of giving us a way to critically measure aspirational missions and grandiose claims. Promises such as "do no evil" and "make the world a better place" are currently little more than marketing vagaries. We ought to be better equipped to bring rigor to these kinds of aspirations, to challenge them effectively, and to encourage integrity. For example, when a company such as Google makes the claim that its technology will make a better world,[9] we should be able to assess this claim in a meaningful way from multiple perspectives, including wellbeing, sustainability and social impact. Positive computing will get us part of the way by allowing us to do so from the perspective of human psychological wellbeing. This approach will provide one piece to the puzzle of proof with regard to whether a technology does indeed deserve to exist.

The Walk-Through

In this book, we hope to support the work of current trailblazers and to facilitate future research and practice by synthesizing multidisciplinary theory, knowledge, and methodologies into a consolidated foundation for a rigorous and prosperous field. In part I, we look at fields outside of computing, such as psychology, economics, and education, as well as at pioneering work within computing that can support or already has begun to address the improvement of wellbeing.

We are privileged to be able to include perspectives from various experts from disciplines such as psychology, neuroscience, and HCI as sidebars throughout the book. Jeremy Bailenson, Timothy Bickmore, danah boyd, Jane Burns, David Caruso, Mihaly Csikszentmihalyi, Felicia Huppert, Mary-Helen Immordino-Yang, Adele Krusche, Jane McGonigal, Jonathan Nicholas, Don Norman, Yvonne Rogers, and J. Mark G. Williams have generously shared aspects of their vision for how future technology might take part in supporting wellbeing.

After a review of the foundational literature, in chapter 5 we propose a theoretical framework and consider appropriate methods for the research and evaluation of positive-computing technologies. We also make efforts to sketch out a scope for the field, looking not only at technologies specially built to support wellbeing, but also at the potential for wellbeing research to enhance the experience of *all* technology.

In part II, we zoom in on a number of specific wellbeing factors as identified in the literature, specifically positive emotions, motivation, engagement, self-awareness, mindfulness, empathy, compassion, and altruism. We look at the literature that correlates these factors to wellbeing, what kinds of strategies exist for fostering them, how technology has already been used to support their development, and possibilities for future work.

Before coming to a close, we take a critical look at issues such as privacy, paternalism, psychological complexity, and autonomy—all of which need to be judiciously explored as part of future work.

Finally, we envision a way forward, including a pragmatic exploration of how current and future work in positive computing might be funded and sustained.

One of the goals of this book is to make a convincing case that considering wellbeing in the design of technology is not only entirely achievable, but also valuable, if not imperative, to building a digital environment that can make a happier and healthier (not just more productive) world. We also hope to show that to enter an age of ubiquitous computing while turning a blind eye to the influence of technology on wellbeing is to accept a kind of convenient ignorance of the real impact of our work and thus to limit our success as designers and developers.

The potential for technology to become a vehicle for worldwide flourishing is huge, and the intentions of enthusiastic professionals are genuine, but in order for our efforts to be effective they must be grounded in evidence and open to evaluation, and, in the end, they must prove themselves. This book attempts to take a first step in what we hope will be an ongoing rigorous and dynamic interdisciplinary journey toward digital experience that is very deeply human centered.

Notes

1. See globalenvision.org/2013/12/18/infographic-there-are-more-mobile-devices-people-world.

2. New ways of structuring a profitable organization around a social benefit come in various forms, including "for-benefit organizations" (e.g., Mozilla), low-profit, limited liability corporations (L3Cs), and "social businesses," proposed by Nobel Peace Prize winner Muhammad Yunus (see yunussb.com or his book *Building Social*

Business). These new organizational models are sometimes described as being part of an emerging "fourth sector" (see fourthsector.net).

3. In 2011, the United Nations officially put happiness on the global agenda, guided by the king of Bhutan's suggestion that "gross national happiness" complement gross national product as an indicator of social progress (see Ryback 2012). Although the current leader of Bhutan has since set aside the idea of gross national happiness, other measures of happiness and life satisfaction have been adopted by policymakers in the United Kingdom, where the National Wellbeing Programme (which carries the slogan "Measuring what matters") was created as part of the Office for National Statistics. The *World Happiness Report* (Helliwell, Layard, & Sachs, 2012) provides a summary of national and international policy initiatives, which we discuss in more detail in chapter 3.

4. In the past decade, psychologists such as Ed Diener, Barbara Fredrickson, Martin Seligman, Sonja Lyubomirsky, and Mihaly Csikszentmihalyi have been part of an effort to extend the focus of psychology and psychiatry beyond a disease model to study the factors of wellbeing and optimal functioning. We discuss positive psychology in chapter 2.

5. Various terms such as *positive technologies*, *positive computing*, and *interaction design for emotional wellbeing* have been used to refer to the potential for technology to support positive psychology and related themes. To our knowledge, it was Tomas Sander who first proposed the term *positive computing* in an article for the edited book *Positive Psychology as Social Change* in 2011. Guiseppe Riva and colleagues use the term *positive technology* in the cyberpsychology context (we look at this work in greater detail in chapter 2).

6. We would include here *The Shallows* by N. Carr, *Alone Together* by S. Turkle, *Nudge* by R. Thaler and C. Sunstein, and also *Flourish* and *Authentic Happiness* by M. Seligman among others.

7. Longitudinal studies by economists show that although wealth has tripled in the United States over the past 30 years, increases in life satisfaction have been marginal. This increase in wealth has likewise come with a significant increase in digital technology use, yet with no significant increase in life satisfaction. Even if we don't expect wellbeing measures to follow Moore's law, a correlation with wellbeing and technology should show more than marginal increases. See Helliwell, Layard, & Sachs 2012 for details.

8. The Australian Quality of Life Centre maintains a useful directory of research instruments. For example, the Personal Wellbeing Index has separate versions for adults, preschoolers, school children, and those with cognitive disabilities (see deakin.edu.au/research/acqol/instruments/instrument.php). You can figure out how you would score on the CESD-R (R = Revised) scale at cesd-r.com.

9. Eric Schmidt, Google's executive chairman, gave a very thorough explanation of this claim in a series of lectures at the University of Cambridge (sms.cam.ac.uk/search?qt_type=sms&qt=Eric+Schmidt&x=16&y=8).

References

Bond, R. M., Fariss, C. J., Jones, J. J., Kramer, A. D. I., Marlow, C., Settle, J. E., … Fowler, J. H. (2012). A 61-million-person experiment in social influence and political mobilization. *Nature*, *489*(7415), 295–298.

Fordyce, M. W. (1977). Development of a program to increase happiness. *Journal of Counseling Psychology*, *24*(6), 511–521.

Harper, R. H. R. (2010). *Texture: Human expression in the age of communications overload*. Cambridge, MA: MIT Press.

Helliwell, J., Layard, R., & Sachs, J. (2012). *World happiness report*. New York: Earth Institute.

Joinson, A., McKenna, K., Postmes, T., & Reips, U.-D. (Eds.). (2007). *The Oxford handbook of Internet psychology* (p. 520). Oxford, UK: Oxford University Press.

Kahneman, D., Krueger, A. B., Schkade, D. A., Schwarz, N. S., & Stone, A. A. (2006). Would you be happier if you were richer? A focusing illusion. *Science*, *312*(5782), 1908–1910.

Ong, A. D., & van Dulmen, M. H. M. (Eds.). (2006). *Oxford handbook of methods in positive psychology*. New York: Oxford University Press.

Ryback, T. W. (2012). The U.N. happiness project. *New York Times*, March 28. At http://nytimes.com/2012/03/29/opinion/the-un-happiness-project.html?pagewanted=all&_r=0.

Schueller, S. M., & Parks, A. C. (2012). Disseminating self-help: Positive psychology exercises in an online trial. *Journal of Medical Internet Research*, *14*(3), e63.

I

2 The Psychology of Wellbeing

"How are you?" "How's it hanging?" "¿Como estás?" "你怎么样?" Humanity's most frequently asked question is none other than an inquiry into another's wellbeing. Responses can vary in sincerity and sophistication: "Good, you?" "Wicked," "Been better," "The clouds of sorrow hang heavy." Despite the variation, we are generally able to understand something of the state of someone's wellbeing following a simple greeting, and, more importantly, we solicit this information before we do anything else. It's not just a social norm—this feedback is vital to any decisions we make about what to do or say next.

Despite its quotidian and timeless nature, this question remains a formidable research question for scientists. Some of the difficulty lies in how science should define and empirically measure variations on "being well." The search for an understanding of happiness and how to attain it is arguably a contender for the world's oldest profession. If we are to look to the academic pursuit of happiness as it has unfolded through time, we find ourselves journeying back at least as far as Aristotle and the Buddha, moving on through various schools of philosophy in Europe, Asia, and the Americas, until we land squarely in the modern world. Today the empirical search for wellbeing rests largely in the hands of psychologists and neuroscientists. Before we can involve digital technology more consciously in this pursuit, we'll need to understand the methods, theory, and practice—as they have been refined over hundreds of years—that have formed our complex modern-day understanding of human psychological wellbeing and its correlates.

This chapter looks at key elements of this understanding from the viewpoints of multiple specializations in psychology and the mind sciences. Needless to say, we could never be anything like comprehensive in one chapter about a subject to which libraries might be devoted, but we do aim to highlight core research and practices that may be particularly helpful to technology researchers and professionals looking to incorporate this knowledge into their practice.

Paradigms of Wellbeing

Because the term *happiness* is so loaded with diverse interpretations (from fleeting hedonic pleasure to consumer spiritualism), scientists refer with greater precision to "optimal human functioning," "optimal mental health," "psychological flourishing," or "psychological wellbeing." It is to psychological wellbeing that we are dedicated in this book (and which we generally shorten simply to "wellbeing"). We occasionally also use the word *flourishing*, which has been widely adopted within the field of positive psychology as a way of emphasizing the optimal (rather than just average) end of possible human psychological functioning.

First off, we should acknowledge that there is an understanding common to all theories of wellbeing that it is contingent on certain basic material needs essential to survival, such as food, water, and shelter. What enhances wellbeing after basic needs are satisfied is more controversial and depends on how wellbeing is defined. For example, is wellbeing defined as the absence of mental dysfunction, in the way that physical health might be described as the absence of illness? Is wellbeing measured as an aggregate of pleasurable experiences (or what percentage of your life you experience positive emotions)? Perhaps it is best understood as the level to which one finds meaning in life and fulfills one's greatest potential. These three perspectives roughly equate with the medical, hedonic, and eudaimonic approaches, which together form the foundations for modern theories of wellbeing. We look at each of these perspectives here.

It's important to note that none of the theories we include herein is simply hypothetical. Each is supported by ample empirical evidence and is associated with a series of measures and validated methodologies for research. The theories don't so much contradict each other as they do focus on different components of wellbeing. For designers of technology, the underlying philosophical standpoint is perhaps less important than the strategies arising from these theories that have been proven to improve wellbeing in practice. We call on examples of these strategies throughout the book.

We believe it would be foolhardy for us to arbitrarily select a theory and posit it as the "right" choice for use in technology fields. Instead, we provide a review geared toward technology designers and imagine that professionals will select (as some already have) a theoretical perspective most appropriate to their context, the backgrounds of their teams, their goals, and their opportunities. The important point is that theory and supporting literature are essential. Work in positive computing might sail aimlessly or, worse, head into harmful waters if not anchored in research-based evidence. Therefore, it's necessary to ground work in existing research, even if the specific literature from which we draw and the disciplinary lens through which we view the problem vary among projects.

For this reason, the framework we propose in chapter 5 is designed to support practitioners in grounding their efforts in the available theory and research, but without prescribing the use of a specific theory. For example, a combination of medical and positive-psychology models of wellbeing shape the work we do with the Young and Well Cooperative Research Centre. A research organization that focuses on the mental health of young people, the center is influenced by the psychologists with whom we work. Specifically, we work to build technologies that support certain psychological strengths such as resilience and autonomy by drawing on the literature in psychiatry and positive psychology. Our target audience contributes via participatory design practice. As new partners get involved in the project, we work with sensitivity to their background and understand that our approaches to influencing and measuring wellbeing may have to adapt over time. Later in the book we look more specifically at how various theories shape design and evaluation in different ways.

The Medical Model—Wellbeing as the Absence of Dysfunction

"How does that make you feel?" asked Sigmund. Despite its wild success as a cliché, if you seek professional assistance for any number of mental health problems, you are more likely to be asked about your appetite, your sleep patterns, and your sense of hopelessness. These questions are just a few in a standard slew that will allow your doctor to determine a diagnosis using a method recognized by the American Psychiatric Association (or the equivalent in your country).

These questions are not random. They have been carefully evaluated in hundreds of studies as accurate indicators of mental illness. Health-care workers, psychiatrists, and insurance companies rely on these methods to determine treatment, write prescriptions, initiate therapy, recommend hospitalization or calculate insurance coverage. The questions included in these standardized questionnaires have been refined over time and after considerable debate have been included in what is known as the *Diagnostic and Statistical Manual* (DSM), recognized by an organization of more than 36,000 American psychiatrists. Similarly, the *International Classification of Diseases* (ICD) is a statistical classification of diseases and related health problems (including mental health) published by the World Health Organization.

Psychiatrists, like other doctors, treat illness, dysfunction, and disease. You can't get much out of a doctor's appointment if there's nothing identifiably wrong with you. You can't, for example, drop in to see your general practitioner because you feel you're not thriving emotionally, you'd like to make wiser decisions, or you want to experience happiness more frequently. The initial evaluation made in the medical field is generally a binary one: you're either sick (and need treatment), or you're not (have a sticker). If you're not ill, your needs will generally fall outside of your doctor's area of professional responsibility.

But the focus of this book is on designing technologies to support and *promote* psychological wellbeing, not specifically for those who are ill and who seek help, but for the population at large, situated as we all are along a continuum from languishing to thriving. Only then can we promote improved life experience and optimum functioning for everyone. Promotion is differentiated from prevention and treatment in the health professions. For example, Mary Ellen O'Connell, Thomas Boat, and Kenneth Warner (2009) describe prevention as the avoidance of risk factors, whereas promotion strives to advance supportive conditions and protective factors. In this context, a medical or psychiatric model may seem inadequate. Nevertheless, even in the context of promotion, a medical model can contribute to our work in many ways.

First, psychiatric methods for diagnosis and intervention have a long history of empirical study and have been extremely successful at evolving diagnosis and treatment for many disorders. Moreover, when we work with teams of mental health professionals, they generally expect to use established medical instruments for assessing the impact of an intervention (even a promotional one). Take, for example, a prototypical randomized control trial evaluating the impact of a preventative intervention on young people at risk of depression (Clarke et al., 2001). In the study, research psychiatrists used cognitive restructuring therapy to prevent the symptoms of depression in young people who were mentally healthy but were nevertheless at risk because their parents were clinically depressed. The study evaluated a preventative intervention using two scales: the CES-D and the DSM-IV Global Assessment of Functioning, which are generally used both before the treatment and again in follow-ups (e.g., 15 months later). Clearly, these scales were especially appropriate in this case because the goal was preventing mental illness. But other studies have used these scales to measure increases in wellbeing by showing decreases in symptoms of depression or anxiety.

Another way to leverage a medical model for informing work in flourishing is to flip it upside down. Felicia Huppert, director of the Cambridge Well-being Institute and wellbeing adviser to the UK government, specializes in multidimensional approaches to the measure of wellbeing. In a recent study, Huppert and her colleague Timothy So (2013) examined various internationally accepted measures of depression, anxiety, and other forms of mental disorder. They aggregated common symptoms from the ICD and DSM (such as hopelessness, lack of interest, and negative emotions) and then looked to their mirror opposites (optimism, engagement, and positive emotions). In this way, they were able to identify two poles that formed ranges both below a tipping point for mental illness as well as above that point into flourishing. In doing so, they identified 10 components of wellbeing, which include competence, emotional stability, engagement, meaning, optimism, positive emotion, positive relationships, resilience, self-esteem, and vitality. These components were analyzed on a sample of 43,000 Europeans from 23 countries, providing national governments and the European Union with a new tool for evaluating progress.

One thing to keep in mind is that if you are designing for people suffering from mental health problems or for those people close to the sufferers, you must ensure that mental health experts are at the helm of your project. There are ethical and legal responsibilities associated with online therapeutic intervention—for example, requirements for allowing people at high risk to connect with professional help directly. For an organizational perspective on creating technologies designed specifically to promote mental health, see the sidebar by Jonathan Nicholas in this chapter.

Although the medical model of mental health will remain essential to work on wellbeing technology, many positive-computing projects will find the model too limited to be useful for promotion rather than treatment. In these cases, researchers often turn to hedonic and eudaimonic models of wellbeing.

Hedonic Psychology—Wellbeing as the Experience of Positive Emotion

When a friend asks, "How are you?" you probably don't base your response on a clinical diagnosis. We're guessing you're more likely to base it on how well your current circumstances match your intentions or simply how well you feel in that moment and have felt of late. If you're stuck working late on your taxes, your answer might be "miserable"; if you're enjoying a nice dinner with friends, it might be "fantastic."

Thinking of wellbeing as something attained through the fulfillment of pleasures has a long history in philosophy. In Greece, Aristippus (born c. 435 BCE) taught that our highest ambition should be to experience as much pleasure as possible. Happiness, he claimed, can be measured as the sum of one's hedonic moments. Many others have written about hedonic pleasures since then. Famously (or infamously), the Marquis De Sade insisted that sensuous pleasures are the ultimate goal in life.

More recently and perhaps more convincingly, Daniel Kahneman, a psychologist and winner of the Nobel Prize in Economics, has looked deeply into the hedonic aspects of human psychology (see, e.g., Kahneman, Diener, & Schwarz, 1999). The field of "hedonic psychology" explores the ancient hedonic view but extends it beyond sensual pleasures to include all the things we judge as pleasant or unpleasant, including the attainment of goals or outcomes in any aspect of life. Kahneman's seminal research in hedonic psychology and behavioral economics has looked at how our quest for pleasurable experiences influences the way we act and think, which in turn manifests in the workings of our economies and societies.

Kahneman's work aggregates both in-the-moment and remembered experiences, and both types of experience have been used from a therapy perspective. In one study (Quoidbach, Berry, Hansenne, & Mikolajczak, 2010), 282 participants completed questionnaires that measured positive affect, life satisfaction, and overall happiness and inquired about their savoring and dampening strategies. Their results showed that mindfulness on the present moment and positive rumination promoted positive emotions and that sharing with others increased measures of life satisfaction. However, mind wandering reduced positive emotions, and ruminating on negative details reduced life satisfaction. Research like this points to the potential importance of remembered experience and focus as a part of technology design.

Modern industrial, architectural, and digital design owes much to a hedonic perspective of wellbeing. Artifacts are designed to heighten feelings of pleasure (at what Don Norman [2005] calls the visceral, behavioral, and reflective levels). User-experience designers seek experiences for their users that are delightful and pleasurable. Researchers in HCI have studied how devices can be built to regulate positive emotions (Hassenzahl & Beu,

2001), and companies such as Apple have built and strengthened a reputation on the idea of product as positive emotional experience.

In his book *The Architecture of Happiness*, philosopher Alain De Botton (2006), describes how art and architecture "talk" to those who experience them and change the way they feel and behave. One could argue that digital technology has the incredibly unique ability to turn the architectural monologue into an interactive dialogue. Digital technologies have the ability also to listen and adapt to what they hear. Imagine an empathic Siri or an emotionally attuned mobile phone. We discuss positive emotions more thoroughly as a factor for increasing wellbeing in chapter 7.

Subjective Wellbeing—If You're Happy and You Know It, Let Us Know

Modern hedonic psychology has come a long way since Aristippus, but there are still problems with relegating evaluations of wellbeing to measures of fleeting emotions, which neglects the long-term overall stability that generally differentiates the concept of wellbeing from definitions of happiness. Kahneman, among others, has approached the need for a measure of longer-lasting characteristics by developing measures of wellbeing based on an individual's self-reported assessment of his or her own life satisfaction. "Subjective wellbeing" (SWB) (Kahneman, Diener, and Schwarz, 1999) consists of the cognitive and affective evaluations of one's life, including life events, life satisfaction, and fulfillment. These measures have been used, for example, for the development of national happiness indices (Diener, 2000; Diener & Suh, 2003), which are increasingly used to inform policymaking in multiple countries (examples are discussed in more detail in chapter 3 from within the multidisciplinary context of economics).

Subjective measures of wellbeing generally consist of three components: *life satisfaction*, the presence of *positive mood*, and the absence of *negative mood*. Life satisfaction is based on more reflective judgment, whereas the latter two refer to hedonic, affective components and can be either retrospective (as in "Over the last week I felt happy") or present focused (as in "I feel happy").

Most of the academic research in hedonic psychology has employed SWB measures that have shown substantial validity, as reflected by their agreement with other types of measures, such as third-party reports and biological measures of wellbeing (e.g., functional magnetic resonance imaging). A review by Ed Diener (2000), for example, highlights what was already known about subjective wellbeing and its different measures at the end of the twentieth century. Progress since then has come on several fronts, including new brain-imaging and genomic techniques (Fredrickson et al., 2013) and digitally facilitated methods for data collection and self-report.

Some research studies employ experience-sampling methods, in which emotions are repeatedly reported at random times during the day (Kahneman, 1999), and others have used diary methods (Bolger, Davis, & Rafaeli, 2003), also common in HCI research, to record memories of good and bad events or satisfaction about different aspects of life. According to these self-reports, people (those not living in extreme poverty or dire circumstances) tend to report being slightly happy. It is uncommon to find people reporting very high or very low levels of wellbeing.

On a time scale, an individual's self-reports can be classified as either "online" (as they occur in real time) or "recalled" (as reported in a diary) or as life evaluations that span long periods of time. These three time scales influence behavior in different ways. For example, Diener has shown that recalled feelings predict future behavior better than moment-to-moment feelings, a finding that relates to how we remember and judge our previous experiences. Because our personal values change very slowly, when we reflect on life satisfaction over a number of weeks or months, our judgments tend to be quite stable. However, reports of satisfaction with life will change over extended periods of time because both personal values and circumstances change more dramatically as we age and as time goes by.

One interesting thing about the effects of external circumstance on wellbeing is our ability to adapt to it. According to the "hedonic treadmill" concept, people adapt to or "get used to" all changes, be they good or bad, by returning to a personal neutral baseline. In other words, that new TV that fills you with happiness the day you buy it will have little to no effect on your happiness level in a month or so. More dramatic is the research showing smaller than expected changes to life satisfaction for both lottery winners and recent paraplegics after their life-changing events (Boyce & Wood, 2011; Brickman, Coates, & Janoff-Bulman, 1978). The hedonic treadmill concept could render efforts to increase happiness pointless if we simply return to a previous set point every time. However, the Boyce and Wood study (2011) shows the importance of personality and attitudes in predicting positive adaptation, and Diener and others have revised the hedonic treadmill model (Diener, Lucas, & Scollon, 2006), arguing that the set point is not neutral, but instead generally positive, and, more importantly, that it can be changed.

Genetic predispositions and environmental influences play out at the cultural level as well. Large-scale longitudinal databases of self-reports allow researchers to compare SWB across cultures and time, noting differences in various dimensions. For example, France is consistently associated with surprisingly low levels of subjective wellbeing, but Scandinavian countries with unusually high levels. Digging deeper, Huppert and So (2013) point out that although France has the highest ranking of all countries on engagement, it has the lowest ranking on self-esteem and is in the bottom for optimism and positive relationships. They highlight this as evidence for why multidimensional measures for wellbeing are critical to understanding differences between people and nations. National measures of wellbeing together with regional and cultural differences represent an ongoing area of study (Diener & Suh, 2003; Huppert et al., 2009). Measures such as the Happy Planet Index, World Happiness Report, and Eurobarometer provide a looking glass into the differences across countries and cultures as well as into the impact of national events and policy interventions. Some of this research is discussed in the next chapter.

Measures of life satisfaction, SWB, and quality of life are all widely used within various economic, social, and research contexts. But positive emotions are only part of the picture. For the rest of it, we turn to Aristotle's notion of eudaimonia.

Eudaimonic Psychology—Wellbeing as Engagement with Meaning and the Fulfillment of Potentials

Few among us eschew pleasure or positive emotion. In fact, most of us spend much of the day seeking pleasures out in small ways, from that nip to the cookie jar or that session of online games to the sitcom after dinner or cuddles before bed. Positive emotions are part of a happy life, but we're nevertheless stuck with the reality that you can get too much of a good thing, and positive emotions alone may not be a complete answer to lasting wellbeing. Here enters the much celebrated notion of the "middle path" or "golden mean," along with theories of wellbeing that go beyond the experience of positive emotion into the realms of engagement, meaning, relationships, and human potential.

Self-Determination Theory—Wellbeing as Determined by Autonomy, Competence, and Relatedness

Don't ask how we can motivate people. That's the wrong question. Ask how we can provide the conditions within which people can motivate themselves.

—Edward Deci, TEDxFlourCity

Richard Ryan and Edward Deci's self-determination theory (SDT),[1] which posits that *autonomy, competence,* and *relatedness* are the key components of both motivation and wellbeing, is one of the theories of wellbeing most readily applied to a technology context, in part because it is relatively straightforward to operationalize.

In order to be self-determined, we must feel autonomous—that is, be able to attribute the outcomes of our activity to our own intentions (what researchers call the "internal locus of causality"). We must feel competence or confident in our ability to meet challenges (e.g., experience optimal challenges and freedom from threats or demeaning evaluations). And finally, we must feel secure and connected to others.

SDT has many implications for design, perhaps the most conspicuous of which is its attention to intrinsic motivation and autonomy. In chapter 7, we look later at how these implications can influence the design of technologies, in particular those that seek to change or support behavior.

Another implication for design stems from the way in which SDT deals with interpersonal, social, and cultural factors. SDT does not suggest that autonomy, competence, and relatedness would be equally *valued* by people from different socioeconomic backgrounds, families, or cultures. It does maintain, however, that environmental conditions that hinder these factors will have negative psychological consequences in *all* social or cultural contexts. According to this line of thought, sociocultural (and, we argue, digital) environments that support these needs can influence wellbeing at both between-person and within-person levels of analysis.

Whereas hedonic theories of wellbeing rely on SWB research, eudaimonic theories often use measures of how well an individual does on a set of factors that support wellbeing (such as autonomy or positive relationships). Those with a eudaimonic perspective have challenged SWB models for being too narrow and a flawed indicator of healthy living. Those with a hedonic perspective, in turn, have argued that eudaimonic criteria are generally defined by experts, whereas the focus on "subjective" in SWB research respects people's individual ideas on what makes a good life. We believe both measures can be valuable to work in technology design, sometimes in combination, and we look more deeply at how wellbeing can be measured from each of these viewpoints in chapter 5.

Combining Hedonic and Eudaimonic Approaches

Many current theories include both hedonic and eudaimonic aspects as factors of wellbeing, such as the model by Huppert and So mentioned previously. Corey Keyes combines emotional wellbeing (hedonic aspects) with aspects of psychological and social wellbeing (eudaimonic) to describe a mental health continuum. Martin Seligman, originator and ongoing champion of the positive-psychology movement, has developed the PERMA model, which stands for Positive Emotions, Engagement, Relationships, Meaning, and Achievement. Seligman and Keyes are among a number of researchers making inroads to our understanding of wellbeing from within the field of positive psychology.

Positive Psychology—Wellbeing as Flourishing

The field of positive psychology at the subjective level is about valued subjective experiences: well-being, contentment, and satisfaction (in the past), hope and optimism (in the future); and flow and happiness (in the present).
—Martin Seligman and Mihaly Csikszentmihalyi, "Positive Psychology"

Thanks to life-saving progress in psychology and psychiatry, many mental disorders can now be diagnosed, treated, and sometimes cured. Psychologists, however, have come to question the nearly exclusive disease focus of their discipline. In 2000, Martin Seligman, then president of the American Psychological Association, and Mihaly Csikszentmihalyi (Seligman & Csikszentmihalyi, 2000) argued for placing greater emphasis on promoting healthy functioning rather than exclusively on dysfunction. The idea of "positive psychology," as they called it, resonated with a great number of researchers and has come to represent an active field of work with ever-increasing influence.

Positive psychology has matured to an extent that it now influences education, policy, management, and mental health. Journals such as the *Journal of Happiness Studies* and the *Journal of Positive Psychology* as well as conferences, symposia, and handbooks of academic literature have developed from this approach. It has been argued that a special term is no longer needed and that a study of healthy and optimal functioning should simply be understood as an essential part of psychology as a whole.

Many researchers in the area of positive psychology have translated their research findings into self-help books for public benefit. These books often have enough detail that they can go some way to informing design work and ideation. They include Seligman's *Flourishing*; Ed Diener and Robert Biswas-Diener's *Happiness*, Barbara Fredrickson's *Positivity*, Sonja Lyubomirsky's *The How of Happiness*, and Daniel Gilbert's *Stumbling on Happiness*, all of which present practical information based on findings from the authors' research.

It's worth noting that some aspects of human behavior and thinking are easier to change than others. Christopher Peterson and Martin Seligman (2004) have identified four sets of components linked to positive mental health: talents, enablers, strengths, and outcomes. Talents are seen as those traits that are hard to change. Enablers include environmental conditions that support wellbeing, such as the right social conditions, a caring family, and so on. Most interesting from a technology perspective are strengths—those personal facets such as curiosity, kindness, and gratitude that are susceptible to interventions and therefore where psychologists and technologists can more easily make a difference.

It was in the context of positive psychology that the terms *positive computing* (Sander, 2011) and *positive technologies* (Botella & Riva, 2012; Riva & Baños, 2012) were first introduced. Tomas Sander's use was hypothetical—a vision and a call to action for technology to play a role in meeting Seligman's challenge that 51 percent of the population be flourishing by 2051. The influential work of Giuseppe Riva, Brenda Wiederhold, Andrea Gaggioli, and their respective teams in cyberpsychology has pioneered research on developing virtual and augmented reality tools for psychotherapy (e.g., Gorini, Gaggioli, Vigna, & Riva, 2008) and for supporting interventions for physical and mental health (e.g., Riva, Cipresso, Mantovani, Dakanalis, & Gaggioli, 2013).

Giuseppe Riva and Rosa Baños (2012) were the first to suggest an approach to supporting the development of tools for positive psychology. They define a "positive technology" approach as "the scientific and applied approach to the use of technology for improving the quality of our personal experience through its structuring, augmentation and replacement." Based on Seligman's original model described in his book *Authentic Happiness* (2002), they divide personal experience into Positive emotions, Engagement, and Meaning (i.e., eudaimonia)—a model that Seligman has now extended to include "Relationships" and "Achievement," or PERMA. Riva and Baños have proposed improving hedonic wellbeing by providing positive emotional and sensorial experiences; improving eudaimonic wellbeing by providing training for "systematic mood induction," "wellbeing," "reminiscence," and "life theme"; and improving social connectedness by providing "shared positive emotional experiences," "wellbeing training" and "setting shared significant goals."

In this book, although we take a deliberately broad and inclusive view of what research can and should inform positive computing, positive psychology models and methods remain some of the most relevant. We have already presented our notion of positive computing from the context of technology in articles for the engineering community (Calvo & Peters, 2013) and within HCI (an early manifestation) (Calvo & Peters, 2012).

The influence of positive psychology can also be found in other related fields, such as emotional intelligence, an area of study that has seen wide real-world application throughout work and education.

Emotional Intelligence

Anyone can become angry—that is easy. But to be angry with the right person, to the right degree, at the right time, for the right purpose, and in the right way—that is not easy.

—G. E. Vaillant, "Positive Mental Health"

A stroll through any bookstore can lead to a generous section on emotional intelligence (EI). Our historically narrow definition of intellectual intelligence was long in need of an upgrade, and incorporating social and emotional capacities has allowed researchers and professionals to better understand some of the sources of success, performance, and even wellbeing separate from intellectual prowess.

In terms of the academic literature, socioemotional intelligence originates with the research of Jack Mayer, David Caruso, and Peter Salovey (Mayer, Caruso, & Salovey, 1999; Salovey & Mayer, 1990) and was popularized and extended by Daniel Goleman (2005). According to Goleman, EI includes the capacities of self-awareness (to recognize your own emotions), self-regulation (to control them), motivation (to have a passion for what you do), empathy (to recognize others emotions), and social skills (to manage relationships with others).

A number of measures of EI have been developed and evaluated. One of the most commonly used is the Mayer-Salovey-Caruso Emotional Intelligence Test (Mayer, Salovey, & Caruso, 2002). Having such measures means that researchers have been able design interventions such as training modules or policies and then evaluate their outcomes.

Hundreds of studies of interventions have been designed to develop EI in businesses, schools, and professional sports. Positive emotions, such as the ones we feel when receiving compliments, tend to increase prosocial behavior. In contrast, empirical evidence has shown that punitive practices can increase antisocial behavior (Mayer, 1995). This research has been used to promote positive interventions rather than punitive ones in schools and prisons. For example, the Los Angeles Unified School District (2007) recently adopted a policy that requires the implementation of systems of positive reinforcement in schools as an alternative to punishment. Interventions like these for social and emotional learning (Payton et al., 2000) are often grounded in EI theory. We discuss EI in greater detail in part II. (For more detail on EI capacities and the potential of technology to support them, see David Caruso's sidebar in chapter 8.)

Buddhist Psychology—a Science of the Mind

The primary orientation of the Buddhist investigative tradition has been toward understanding the human mind and its various functions. The assumption here is that by gaining deeper insight into the human psyche, we might find ways of transforming our thoughts, emotions and their underlying propensities so that a more wholesome and fulfilling way of being can be found. It is in this context that the Buddhist tradition has devised a rich classification of mental states, as well as contemplative techniques for refining specific mental qualities.

—His Holiness the Dalai Lama, "Science at the Crossroads"

To focus only on Western theory would be strangely remiss for a topic such as wellbeing, which has been studied systematically by Eastern philosophers for thousands of years. An interest in Buddhism for its practices such as mindfulness and meditation and for the culture of peace and compassion it represents has led to a growing integration of Buddhist philosophy into Western notions of wellbeing.

This has been possible in part because Buddhist philosophy and practice can be investigated separate from the religious aspects of cultural rituals and belief systems in which it is nested. It is probably the relatively barebones, nonreligious style of Zen that has made it one of the most widely accessed sources of Buddhist thought in Western culture. In fact, the term *zen* has become a clichéd colloquial synonym for simplicity and tranquility of mind (visit any home furnishings store for evidence). But the other essential element that makes Buddhist philosophy amenable to a partnership with Western science is its commitment to empiricism.

The Dalai Lama is adamant that Buddhist doctrine is subject to scientific evaluation and should change in light of new evidence. He explains that "in the Buddhist investigative tradition, between the three recognized sources of knowledge—experience, reason and testimony—it is the evidence of the experience that takes precedence, with reason coming second and testimony last. This means that, in the Buddhist investigation of reality, at least in principle, empirical evidence should triumph over scriptural authority, no matter how deeply venerated a scripture may be" (2005).

Psychologist Paul Gilbert (2011) puts it this way:

For thousands of years Buddhist scholars and devotees studied and developed practices of introspective and reflective psychology and an ethic based on compassionate insights—these are ways by which individuals can become very familiar with their minds, learn to stabilise and organise them for their wellbeing, and cultivate key qualities that are associated with personal and social health. ... While the focus of science has been on understanding and alleviating the physical nature and causes of pain, spiritual traditions like Buddhism have tended to focus more on alleviating suffering, that is, working with how the mind reacts to pain.

It is perhaps owing to this empirical stance that technologists interested in Buddhist philosophy are now significant enough in number to have motivated the creation of the annual "Buddhist Geeks" conference. Interest lies in how Buddhist practitioners investigate the interrelationships between emotion, cognition, and behavior as well as in Buddhist practices such as mindfulness and various forms of meditation, taught for centuries as paths to wellbeing. These practices have also of late been increasingly evaluated by Western psychology and neuroscience. Work in multiple fields using multiple measures has consistently shown Buddhist meditative and mind-training practices to be highly effective for treating mental illness and increasing wellbeing. So compelling is the evidence of their effectiveness that mental health professionals at institutions such as Oxford, Harvard, Stanford, Yale, the University of California at Berkeley, and the University of Wisconsin at Madison, among many other institutions around the world, have incorporated them into clinical and research work.

Jon Kabat-Zinn (2003), the originator of the Mindfulness-Based Stress Reduction program, one of the most successful integrations of Buddhist and Western psychology to date (which we discuss in greater detail in chapter 9), adeptly describes Buddhist practices in untraditional and elucidating terms: "Of course, the Buddha himself was not a Buddhist. One might think of dharma as a sort of universal generative grammar, an innate set of empirically testable rules that govern and describe the generation of the inward, first-person experiences of suffering and happiness in human beings. ... It is neither a belief, an ideology, nor a philosophy. Rather, it is a coherent phenomenological description of the nature of mind, emotion, and suffering and its potential release, based on highly refined practices aimed at systematically training and cultivating various aspects of mind and heart via the faculty of mindful attention." In the following chapters of this book, we look more closely at how these practices are used in various contexts.

Biology and Neuroscience—Wellbeing as Physiologically Identifiable

Researchers in biology and neuroscience have used physiological and brain signals to detect and understand individual emotions. Others study biological factors that influence wellbeing (such as genes or physical health), while some investigate how those biological systems interact with environmental conditions.

This work intersects with HCI most clearly in affective computing. Rafael's research group has been among those to use physiological signals to detect emotions during HCI, particularly for applications within education and mental health. For example, physiological signals can be used to measure the impact of feedback when students receive it during online activities (Pour, Hussain, AlZoubi, D'Mello, & Calvo, 2010). Moreover, signals from multiple physiological systems can be combined, including electroencephalography (EEG), electromyography, skin conductivity, and respiration (AlZoubi, Hussain, D'Mello, & Calvo, 2011). We come back to this work with respect to affective computing in chapter 4.

Neuroscience researchers seek to identify patterns of electrical and chemical activity in the brain that correlate with the emotion, cognition, and behavior we experience. In the past two decades, their research has come to include positive emotions as well as characteristics associated with increases in wellbeing, such as resilience and meditative practice.

Using brain-imaging techniques, scientists can learn more about the brain's structures and the processes behind emotions. For example, researchers may have found the neural network responsible for answering our opening question: "How are you?" The anterior insula cortex seems to contain the interoceptive representation of our embodied feelings and emotions (Craig, 2009b). Together with the anterior cingulate cortex, it is activated in subjects experiencing emotional feelings such as love, anger, fear, sadness, happiness, indignation, social exclusion, and empathy.

These neural correlates have been used to propose a model of awareness that includes homeostatic, environmental, hedonic, motivational, social, and cognitive activity to describe both a "global emotional moment" and the fact that a series of such moments produces a representation of sentiments over time. Eight prosocial positive emotions (love, hope, joy, forgiveness, compassion, faith, awe, and gratitude) are often identified as the components of wellbeing in this model. Notably, almost all involve human connection rather than just the self. These models do not require that all our emotions be positive and acknowledge that negative emotions are necessary for survival (Craig, 2009a, 2009b).

Affective and social neuroscience recognize that our brain is also shaped by what we experience. For example, studies using functional magnetic resonance imaging (fMRI) show that early stressful and nurturing environments have a strong effect on how the brain develops. Richard Davidson and others (e.g., Davidson & McEwen, 2012) have been gathering evidence that certain interventions can be intentionally designed to promote prosocial behavior and wellbeing. According to Davidson's research, structural changes in the brain can be triggered by regular exercise, cognitive therapy, and meditation practices, suggesting that we can develop training practices for this purpose. This work poses tantalizing questions for the potential impacts of technologies in these same areas.

In a recent article, Davidson and colleagues (2012) discuss how such results can inform education. They posit that it should be possible to support prosocial behaviors and academic success in young people by developing the underlying elements of wellbeing through systematic contemplative practices that have been shown to be effective and to trigger neuroplastic change. They have also pointed to the potential for technologies such as videogames to be used to develop positive characteristics, including mindfulness and empathy.

Others who study the biological factors of wellbeing look at the relationship of physical behaviors such as circadian rhythms, diet, and exercise to psychological health. For example, Ian Hickie at the University of Sydney studies the chronobiology system (our physiological clock) and its effect on depression. Even research in this area can inform work in positive computing. For example, together with Hickie we are exploring how information about sleep cycles collected from social networks might be used for detecting people at risk of depression.

Personality traits (Costa & McCrae, 1992) and genetics are other acknowledged determinants of wellbeing. During the 1990s, neuroscientists hoped to be close to identifying the genetic determinants of mental illnesses. Since then we have come to better recognize the complexity and sheer number of genes involved in both mental disorders and in flourishing, yet progress has been made on many fronts. A groundbreaking paper (Caspi et al., 2003), for example, revealed the impact of a certain gene configuration known as the 5-HTT promoter that determines how well our neurons transport serotonin—a neurotransmitter famously linked to depression and wellbeing. They found that those with one or two copies of the short allele of the gene were more vulnerable to depression when faced with life-stressing situations. Another study (Pluess & Belsky, 2013) resulted in four categories of resilience: (1) those that are highly reactive to both negative and positive events, (2) those that are low reactive to both types of events, (3) those that are vulnerable to negative events (low resilience), and (4) those that are more influenced by positive events or "vantage sensitive."

Even the apparently predetermined factors of genetics and personality traits can be influenced and changed. For instance, we now understand that gene expressions are modified by the environment and personal experience, an area of research known as "epigenetics." One extraordinary example is presented in recent work by Barbara Fredrickson and her colleagues that shows how different forms of wellbeing correlate with different gene transcription (as discussed in the next section).

Hedonic versus Eudaimonic Wellbeing at the Molecular Level

If you're confused about whether to take a hedonic or eudaimonic approach to wellbeing, you might consider letting your cells decide. Fascinating new research (Fredrickson et al., 2013) suggests that the human genome may be more sensitive to the differences between hedonic and eudaimonic wellbeing than either our affective states or our philosophers have been. It turns out that hedonic wellbeing and eudaimonic wellbeing are correlated with different patterns of gene expression. Moreover, the molecular patterns associated with hedonic wellbeing are associated with a stress response that promotes inflammation and decreases antibody production. In contrast, eudaimonic wellbeing is associated with transcription patterns that *increase* antibody production. Fredrickson and her colleagues conclude: "If 'the good life' is a long and healthy life free from the allostatic load of chronic stress, threat, and uncertainty, CTRA gene expression may provide a negative reference point for how not to live. ... If we ask which type of happiness most directly opposes that molecular antipode, a functional genomic perspective favors eudaimonia". According to their findings, hedonic forms of wellbeing (arising from pleasure) are associated with increases in a particular type of stress-related gene expression, whereas eudaimonic wellbeing (arising from connectedness and purpose) is associated with decreases in the same stress-related gene expression. The message is that in the long term it's eudaimonic wellbeing that promotes health.

The growth in scientific and popular understanding of wellbeing over the past decade has been transformative, but how exactly does one draw a connection between theories of wellbeing and the design of technology? In our experience, the philosophical underpinnings of any design work, be they explicit or unconscious, can have profound effects on design outcomes. By way of demonstrating just how different wellbeing theories might support different and often complementary design approaches, we take a look at the driverless car in the hypothetical case study in the next section.

Wellbeing-Informed Design—a Hypothetical Case Study

The challenges of designing driverless cars go far beyond aerodynamics, fuel efficiency, and safety. We speculate here (albeit playfully) about the approach that four different designers, drawing on four different theoretical viewpoints, might take to the design process.

- **The hedonic designer:** *Driving time is pleasure time.* The hedonic designer will focus on improving those aspects of car design from which we derive positive emotion (heated leather seats, movie screens, and the flower vase in a Volkswagen Beetle come to mind). This designer might improve the quality of the sound system by personalizing music to mood or include massage functions in the chair (we're sold already). She might also seek to identify and remove aspects that caused negative emotions in previous models. She could look to develop driver happiness over a longer term as well—for example, by providing information to help the driver understand his own triggers for road rage. Like an emotionally aware geographical positioning system (GPS), the hedonic car would choose routes based on the feelings of happiness they inspire—for example, by avoiding traffic or privileging scenic views.
- **The SDT designer:** *Driving is about connectedness, competence, and control.* A designer guided by SDT would be concerned with a user's sense of autonomy, competence, and relatedness. As such, he might seek to compensate the sense of autonomy lost by not physically doing the driving. Features that allow the user to shift to manual control, to enact some steps modularly, and/or to give clear directions to the system would be critical. Research on brain–computer interfaces and wheelchair design for

paraplegics make the human need for autonomy very clear (Nijholt et al., 2008). A nondriving driver's sense of autonomy relies on the driver's sense of competence in controlling the vehicle, in endorsing its behavior, and in getting where he wants to go. Finally, the SDT designer may look to support a feeling of relatedness via easy communication with the driver's network of contacts or perhaps with other people in nearby cars or the people and places going past.

- **The values-sensitive designer:** *Driving the way you think it should be.* Whereas vehicle engineers may value automation for its own sake and without question, those who love to drive may consider automation an absolute killjoy. A designer with a background in values-sensitive design (VSD) will seek to make explicit particular values, relating both to the audience and to the designers of the vehicle. To those considering overall effects on society, automation is a threat to the livelihood of the millions of people who make a living from driving. A VS designer might also focus on the serious issues of privacy posed by a driverless car, which may be of greater or lesser importance to different cultural groups. For example, since a driverless car is connected to a mapping system that tracks its coordinates, a VS designer might aim to add a way for GPS data to be anonymized, encrypted, or customized. Although VSD is not a theory of wellbeing or based on a theory of wellbeing, work on values will inevitably interact with work on wellbeing (we discuss this area in more detail in chapter 4).

- **The biological wellbeing designer:** *Driving that's good for you.* A designer aware of the relationships between physical health and psychological wellbeing might seek to make better use of the copious amount of time a user spends sitting down in a car. He might incorporate exercise devices into the seating (modeling her design after the Flintstones' car for example), design pods for power napping, provide a fresh-water dispenser, or program the car to deliberately park several blocks away to encourage the user to walk.

In this chapter, we have endeavored to highlight and synthesize a number of growing research areas that inform the science of optimal human functioning. We have also sought to demonstrate how various theories of wellbeing can practically influence design decisions and technological affordances. Although the focus has been on the physical and mental health fields, in the next chapter we expand our horizon to include some of the critical fonts of discovery emerging from disciplines outside of health—from economics and policy to architecture and education.

Expert Perspectives—Technology for Mental Health

Inspiring Projects—Opportunities for Mental Health and Technology

Jonathan Nicholas, Inspire Foundation

In 1998, Inspire launched the world's first online mental health service—ReachOut.com. Since that time, technology has transformed many aspects of our lives, from business to entertainment to how we connect to others. The potential for Internet and mobile technology to similarly transform mental health and wellbeing is enormous—particularly through the provision of targeted and scalable services that enable people to manage their own health. Through their ability to automate processes and scale efficiently, technology-based services can cast a wider net, simultaneously helping larger numbers of people and doing so more affordably. The result is a twenty-first-century model of mental health care that integrates traditional services, such as counseling, with scalable services that allow people to monitor, manage, and improve their own mental health. The goal should be to enable all people to access the right help at the right time in the way they want it.

Our experience of delivering ReachOut.com in Australia, Ireland, and the United States has provided some insight into how this might occur. ReachOut.com reaches 1.6 million unique visitors each year in Australia alone and has the potential to reach many more and for considerably lower cost than traditional commercial and government mental health services.

One of our biggest challenges in reaching this goal of a twenty-first-century mental health system will be to ensure that the user is placed at the

(continued)

> center of that system and to build that system around mental health promotion. We can achieve these things by better integrating technology that enables people to manage and monitor their own wellbeing and assist them with evidence-based advice for personalized mental health care. As one of the pioneers in e–mental health, we are committed to this technology.
>
> We recognize that we can't do it alone, however, and need to form partnerships with researchers and policymakers to build the evidence for these new services and then take them to scale. One of the challenges we continue to face is that "traditional" research processes are often unsuitable in a context where producing innovative and relevant services relies on much quicker timeframes. In this sense, we sincerely welcome positive-computing initiatives that center technology research and practice on mental health and wellbeing support.
>
> Our experience of delivering e–mental health services for more than 15 years is that technology continues to provide exciting opportunities to improve and promote mental health. Taking advantage of these opportunities will require a commitment to research and collaboration between technical and clinical professionals and ultimately a commitment to developing a twenty-first-century mental health system that will enable all people to thrive.

(continued)

Measuring Subjective Wellbeing

Felicia Huppert, University of Cambridge

There is an increasing interest among researchers, organizations, and governments in measuring subjective wellbeing. We need to ask why this is, how it can be done, and what exactly should be measured. The "why" stems from the recognition that wellbeing arises from how we *experience* our lives, not the mere objective facts of our lives—such as our income, job, health, housing, and so on. There is evidence that people with high levels of wellbeing are healthier, more productive, and more creative and have better relationships with others, so high subjective wellbeing is a desirable goal for individuals and society alike.

How can subjective wellbeing be measured? Skeptics sometimes say that subjective experiences such as happiness cannot in principle be measured. Yet most of us are able to indicate how much we enjoyed a meal or a movie or rate our level of pain on a scale from 0 to 10 when asked by a doctor. Likewise, it is widely accepted that individuals can reliably rate symptoms of distress, such as sadness or anxiety, so there is no reason to suppose they cannot also reliably rate positive experiences. Perhaps more compellingly, many studies show that verbal reports of positive experiences such as happiness or interest are highly correlated with objective measures such as facial expression.

(continued)

> The field of subjective wellbeing has also received a great boost from neuroscience because it can be demonstrated that when people report particular experiences, there are patterns of brain activation in regions known to be involved in the neural pathways associated with such experiences.
>
> Since it is important to measure subjective wellbeing, and it is clear that it can be done, we need to consider exactly what should be measured. Studies have traditionally used generic single-item questions about happiness or life satisfaction. But wellbeing is more than a positive feeling or a positive life evaluation. It involves both feeling good and functioning effectively. Feeling and functioning can be measured using questions with different timeframes, including ongoing experiences, recent experiences, and typical experiences.
>
> Importantly, wellbeing is a multidimensional construct that includes feelings, evaluations, and perceptions of how well a person is functioning across different aspects of his or her life. Scholars may differ in what they regard as the central components of wellbeing, but there is consensus about its multidimensional nature. In an empirically derived approach, the components of positive wellbeing (or flourishing) have been conceptualized as the opposite of the symptoms of ill-being—that is, the common mental disorders, namely depression and anxiety. This conceptualization has led to the identification of ten features of flourishing: positive emotion, engagement, meaning, self-esteem, optimism, vitality, resilience, sense of competence, emotional stability, and positive relationships. Measuring multiple features of wellbeing in this way has allowed the discovery of major group and cross-national differences in wellbeing profiles. Future research using this approach can elucidate which features are affected by specific interventions or policies.
>
> As distinguished economist Gus O'Donnell states in relation to wellbeing, "If you treasure it, measure it."

Note

1. See http://www.selfdeterminationtheory.org/.

References

AlZoubi, O., Hussain, M. S., D'Mello, S., & Calvo, R. A. (2011). Affective modeling from multichannel physiology: Analysis of day differences. In *Affective computing and intelligent interaction: International conference, ACII 2011, proceedings* (pp. 4–13). Heidelberg: Springer-Verlag Berlin.

Bolger, N., Davis, A., & Rafaeli, E. (2003). Diary methods: Capturing life as it is lived. *Annual Review of Psychology, 54*(1), 579–616.

Botella, C., & Riva, G. (2012). The present and future of positive technologies. *Cyberpsychology, Behavior, and Social Networking, 15*(2), 78–84.

Boyce, C. J., & Wood, A. M. (2011). Personality prior to disability determines adaptation: Agreeable individuals recover lost life satisfaction faster and more completely. *Psychological Science, 22*(11), 1397–1402.

Brickman, P., Coates, D., & Janoff-Bulman, R. (1978). Lottery winners and accident victims: Is happiness relative? *Journal of Personality and Social Psychology, 36*(8), 917–927.

Calvo, R. A., & Peters, D. (2012). Positive computing: technology for a wiser world. *Interactions (New York, N.Y.), 19*(2), 28–31.

Calvo, R. A., & Peters, D. (2013). Promoting psychological wellbeing: loftier goals for new technologies. *IEEE Technology and Society, 32*(4), 19–21.

Caspi, A., Sugden, K., Moffitt, T. E., Taylor, A., Craig, I. W., Harrington, H., ... Poulton, R. (2003). Influence of life stress on depression: Moderation by a polymorphism in the 5-HTT gene. *Science Signaling, 301*(5631), 386–389.

Clarke, G. N., Hornbrook, M., Lynch, F., Polen, M., Gale, J., Beardslee, W., ... Seeley, J. (2001). A randomized trial of a group cognitive intervention for preventing depression in adolescent offspring of depressed parents. *Archives of General Psychiatry, 58*(12), 1127–1134.

Costa, P., & McCrae, R. (1992). The five-factor model of personality and its relevance to personality disorders. *Journal of Personality Disorders, 6*, 343–359.

Craig, A. D. B. (2009a). Emotional moments across time: A possible neural basis for time perception in the anterior insula. *Philosophical Transactions B, 364*(1525), 1933–1942.

Craig, A. D. B. (2009b). How do you feel—now? The anterior insula and human awareness. *Nature Reviews, Neuroscience, 10*(1), 59–70.

Davidson, R., Dunne, J., Eccles, J. S., Engle, A., Greenberg, M., Jennings, P., ... Vago, D. (2012). Contemplative practices and mental training: Prospects for American education. *Child Development Perspectives, 6*(2), 146–153.

Davidson, R. J., & McEwen, B. S. (2012). Social influences on neuroplasticity: Stress and interventions to promote well-being. *Nature Neuroscience, 15,* 689–695. doi:10.1038/nn.3093.

De Botton, A. (2006). *The Architecture of Happiness.* New York: Pantheon.

Diener, E. (2000). Subjective well-being: The science of happiness and a proposal for a national index. *American Psychologist, 55*(1), 34–43.

Diener, E., Lucas, R. E., & Scollon, C. N. (2006). Beyond the hedonic treadmill: Revising the adaptation theory of well-being. *American Psychologist, 61*(4), 305.

Diener, E., & Suh, E. M. (2003). National differences in subjective well-being. In *Well-being: The foundations of hedonic psychology* (pp. 434–450). New York: Russell Sage Foundation.

Fredrickson, B. L., Grewen, K. M., Coffey, K. A., Algoe, S. B., Firestine, A. M., & Arevalo, J. M. G. ... Cole, S. W. (2013). A functional genomic perspective on human well-being. *Proceedings of the National Academy of Sciences of the United States of America, 110*(33), 13684–13689.

Gilbert, P., & Choden (2013). *Mindful compassion.* London: Constable & Robinson.

Goleman, D. (2005). *Emotional intelligence: Why it can matter more than IQ.* 10th anniversary ed. New York: Bantam Books.

Gorini, A., Gaggioli, A., Vigna, C., & Riva, G. (2008). A second life for eHealth: Prospects for the use of 3-D virtual worlds in clinical psychology. *Journal of Medical Internet Research, 10*(3), e21.

Hassenzahl, M., & Beu, A. (2001). Engineering joy. *IEEE Software, 18*(February), 70–76.

His Holiness the 14th Dalai Lama of Tibet. (2005). "Science at the crossroads." November 12. At http://dalailama.com/messages/buddhism/science-at-the-crossroads.

Huppert, F., Marks, N., Clark, A., Siegrist, J., Stutzer, A., Vittersø, J., ... Wahrendorf, M. (2009). Measuring well-being across Europe: Description of the ESS well-being module and preliminary findings. *Social Indicators Research, 91*(3), 301–315.

Huppert, F. A., & So, T. T. C. (2013). Flourishing across Europe: Application of a new conceptual framework for defining well-being. *Social Indicators Research, 110*(3), 837–861.

Kabat-Zinn, J. (2003). Mindfulness-based interventions in context: Past, present, and future. *Clinical Psychology: Science and Practice, 10*(2), 144–156.

Kahneman, D. (1999). Objective happiness. In D. Kahneman, E. Diener, & N. Schwarz (Eds.), *Well-being: The foundations of hedonic psychology* (pp. 3–25). New York: Russell Sage Foundation.

Kahneman, D., Diener, E., & Schwarz, N. (Eds.). (1999). *Well-being: The foundations of hedonic psychology*. New York: Russell Sage Foundation.

Mayer, G. R. (1995). Preventing antisocial behavior in the schools. *Journal of Applied Behavior Analysis, 28*(4), 467–478.

Mayer, J. D., Caruso, D. R., & Salovey, P. (1999). Emotional intelligence meets traditional standards for an intelligence. *Intelligence, 27*(4), 267–298.

Mayer, J., Salovey, P., & Caruso, D. (2002). *Emotional intelligence test*. Toronto: MSCEIT.

Nijholt, A., Tan, D., Pfurtscheller, G., Brunner, C., Millán, J. R., Allison, B., et al. (2008). Brain–computer interfacing for intelligent systems. *IEEE Intelligent Systems, 23*(3), 72–79.

Norman, D. A. (2005). *Emotional design: Why we love (or hate) everyday things*. New York: Basic Books.

O'Connell, M. E., Boat, T., & Warner, K. (2009). *Preventing mental, emotional, and behavioral disorders among young people: Progress and possibilities*. Washington, DC: National Academies Press.

Payton, J., Wardlaw, D., Graczyk, P. A., Bloodworth, M. R., Tompsett, C. J., & Weissberg, R. P. (2000). Social and emotional learning: A framework for promoting mental health and reducing risk behavior in children and youth. *Journal of School Health, 70*(5), 179–185.

Peterson, C., & Seligman, M. E. P. (2004). *Character strengths and virtues: A handbook and classification* (p. 800). Oxford: Oxford University Press.

Pluess, M., & Belsky, J. (2013). Vantage sensitivity: Individual differences in response to positive experiences. *Psychological Bulletin, 139*(4), 901–916. doi: 10.1037/a0030196.

Pour, P. A., Hussain, M. S., AlZoubi, O., D'Mello, S. K., & Calvo, R. A. (2010). The impact of system feedback on learners' affective and physiological states. In V. Aleven, J. Kay, & J. Mostow (Eds.), *Intelligent tutoring systems* (vol. 6094, pp. 264–273). Berlin: Springer.

Quoidbach, J., Berry, E. V., Hansenne, M., & Mikolajczak, M. (2010). Positive emotion regulation and well-being: Comparing the impact of eight savoring and dampening strategies. *Personality and Individual Differences*, *49*(5), 368–373.

Riva, G., & Baños, R. (2012). Positive technology: Using interactive technologies to promote positive functioning. *Cyberpsychology, Behavior, and Social Networking*, *15*(2), 69–77.

Riva, G., Cipresso, P., Mantovani, F., Dakanalis, A., & Gaggioli, A. (2013). *New technologies for improving the psychological treatment of obesity*. Berlin: Springer.

Salovey, P., & Mayer, J. D. (1990). Emotional intelligence. *Imagination, Cognition, and Personality*, *9*(3), 185–211.

Sander, T. (2011). Positive computing. In R. Biswas-Diener (Ed.), *Positive psychology as social change* (pp. 309–326). New York: Springer.

Seligman, M. (2002). *Authentic happiness: Using the new positive psychology to realize your potential for lasting fulfillment*. New York: Free Press.

Seligman, M. E. P., & Csikszentmihalyi, M. (2000). Positive psychology: An introduction. *American Psychologist*, *55*(1), 5.

Vaillant, G. E. (2012). Positive mental health: Is there a cross-cultural definition? *World Psychiatry: Official Journal of the World Psychiatric Association (WPA)*, *11*(2), 93–99.

3 Multidisciplinary Foundations

Let's be honest: engineers and computer scientists aren't known for advanced social skills or for their perspicacity with regard to human emotions. Some would also be quick to point out we have a weakness for letting technological interests drive all our decision making. But shameless stereotyping aside, we believe that when it comes to designing for wellbeing, no matter what field you're in, it's critical not to go it alone.

No attempt to influence or investigate issues as multifaceted as human psychological wellbeing should be undertaken without a rigorously multidisciplinary approach. Neglecting human experience is bad for technology design generally, but it is totally counterproductive for positive computing. Truly human-centered design for positive computing will rely on interdisciplinary teams and collaboration.

In addition to work in psychology, medicine, and brain science, our contemporary understanding of wellbeing has been contingent on progress in anthropology, sociology, philosophy, economics, public policy, media studies, and even architecture, literature, and art. In this chapter, we look at wellbeing through the lenses of some of these disciplines and at how they have charted significant territory on the journey to understanding psychological flourishing.

Economics—Wellbeing as Something Money Still Can't Buy

Economics is surprisingly, or perhaps quite logically, one of the richest sources of research on wellbeing. Most notably, it has provided rigorous methods for measuring levels of wellbeing across populations. For example, the *Handbook on the Economics of Happiness* (Bruni & Porta, 2007) is a compendium of research at the intersection of economics, public policy, and psychological wellbeing. *Well-Being: The Foundations of Hedonic Psychology* (Kahneman, Diener, & Schwarz, 1999) is an earlier volume that supported much of the growth in this field. In the chapter "Objective Happiness," Daniel Kahneman provides one of the best introductions to wellbeing measurement techniques available, a topic we discuss in greater detail in chapter 5 on methodologies.

Economics may also help to explain the recent growth of lay interest in wellbeing psychology. Richard Ryan and Edward Deci (2001) have identified two periods of peaking interest in wellbeing in the history of psychology: the 1960s and the 2000s. They note that these two periods coincide with times of affluence.

Despite the recent global financial crisis, those in industrialized Western society are still in general relatively wealthier than we ever have been before—gross national product per capita has tripled since the 1960s (Helliwell, Layard, & Sachs, 2012). This wealth has also come with great progress in the development of digital technologies—we are certainly more immersed in technology than we ever have been before. And yet for the so-called industrialized world neither increased technification nor continued increases in wealth have led to much greater psychological wellbeing according to the statistics.

A seminal study by Richard Easterlin (1974) famously provided early evidence that the link between wealth and happiness is weaker than popular culture would have us think. Easterlin found that wealthier people within a country do tend to be more satisfied with their lives at any given point in time, but that happiness does not increase with economic growth over time (a finding henceforth dubbed the "Easterlin Paradox"). In other words, the increased happiness that money can provide is *relative* and seems to adapt as economies grow. Therefore, using economic growth measures, such as gross domestic product, as proxy measures for a nation's wellbeing is ineffective.

It's also worth noting that, according to Easterlin's research, the impact that wealth has on wellbeing is significant up to the point at which basic needs are met. After that point, the impact is small. Easterlin's results suggest that relative increases in happiness are influenced not only by an individual's state of wealth, but also by how it compares to others. The powerful impact of comparison on wellbeing is echoed in psychological research on happiness and self-esteem, which we look at in more detail in part II.

As Easterlin himself noted, his original work had some limitations. The study combined data from 29 Gallup Poll–type surveys and produced a single measure of self-reported wellbeing on a scale from 1 to 10. Forty years later, approaches to measuring wellbeing have evolved significantly.

Researchers such as Felicia Huppert and Timothy So have studied such measures seeking multidimensional approaches for greater validity and explicatory power. We look more closely at these measures in the methodologies chapter. But for now we consider how findings in economics have influenced change in public policy.

Easterlin's work has been the focus of dozens of studies and remains hotly debated. The United Nations' first *World Happiness Report* (Helliwell et al., 2012) provides a summary of the sometimes contradictory literature on this topic. In general, it is understood that economic growth does not automatically increase the average wellbeing of a population. But both wellbeing and economic status need to be unpacked further. For example, the impact of *income comparison* is significant. When people are asked, "How important is it for you to compare your income with other people's income?" the greater the importance they report, the less satisfied they are with their lives (based on the European Social Survey, as cited in Helliwell et al., 2012).

Furthermore, employment (and presumably its concomitant benefits, security and self-esteem) is a core component of the economic equation. There is no doubt that higher levels of *employment* have a positive effect on wellbeing. But economic booms often produce inflation, which has a negative effect. In the end, stability seems to be a good target, especially considering evidence that loss aversion (losing x dollars) has a bigger impact than an equivalent gain (receiving x dollars).

It's hard to avoid speculation that the growing interest in wellbeing in the Western world is at least in part due to a gradual societal realization that money doesn't, after all, bring reliable happiness. But this is probably not the whole picture. Significant advancements in our ability to research, evaluate, and operationalize wellbeing are also at the heart of this trend.

For instance, after a seminal study published in *Science* (Golder & Macy, 2011) showed that social media data could be used to study moods over time, many other studies plundering the wealth of publically available social interaction data have followed in aid of better understanding the human experience. One such study (Mitchell, Harris, Frank, Dodds, & Danforth, 2013) combined geotagged tweets (80 million words in total) with demographic and health information from annual surveys. They used the data to build taxonomies that describe the happiness of states and cities across the United States, to correlate demographic information with wellbeing, and to correlate linguistic features to levels of education and even obesity rates. These studies provide evidence that public social media data can be used to investigate communities' overall wellbeing levels in real time—an incredible opportunity for nonintrusive methods to inform research and even policy.

Government and Policy: Increasing Gross National Happiness and General Wellbeing

The care of human life and happiness, and not their destruction, is the sole legitimate object of government.
—Thomas Jefferson, To the Citizens of Washington County, Maryland (1809)

In an influential paper, psychologists Ed Diener and Martin Seligman (2004) argued that "policy decisions at the organizational, corporate and governmental levels should be more heavily influenced by issues related to wellbeing—people's evaluations and feelings about their lives." They proposed the creation of a national wellbeing index that would periodically measure wellbeing in representative samples of the population. The index would be multidimensional, including "positive and negative emotions, engagement, purpose and meaning, optimism and trust, and the broad construct of life satisfaction," and would be assessed and updated periodically so it could more accurately inform policymaking. Since then, much has happened to see Deiner and Seligman's vision made real.

In 2007, a group of partners including the European Commission, the European Parliament, and the Organization for Economic Cooperation and Development hosted a high-level conference called "Beyond GDP," with the objectives of determining "which indices are most appropriate to measure progress, and how these can best be integrated into the decision-making process and taken up by public debate." The partners continue to work on developing and measuring social, environmental, and wellbeing indicators.[1]

In 2008, French president Nicolas Sarkozy commissioned a panel of experts, including Nobel Prize–winning economists Joseph Stiglitz and Amartya Sen, to reassess measures of national progress. The resulting report moved the French government to launch a new era in which national measures of progress would take wellbeing into account.

In 2011, the United Kingdom launched the National Well-Being Programme, a part of the Office for National Statistics, with the motto "Measuring what matters." The New Economics Foundation (whose tagline is "economics as if people and the planet mattered") created the Happy Planet Index, which combines data on experienced wellbeing, life expectancy, and ecological footprint into a "global measure of sustainable well-being."[2]

Although the United States has been slower to consider measures beyond gross domestic product, individual cities and counties have established regional measurement initiatives to inform local leadership, and the federal government established a panel to investigate measures of happiness. US commercial initiatives are also collecting data in ways similar to government efforts in other countries: The Gallup–Healthways Well-Being Index undertakes an impressive live daily assessment of health and wellbeing measures across the United States.[3]

But policy change has by no means been relegated to Europe and North America. In fact, the seeds of this movement were planted by the king of Bhutan, who in 1972 declared that gross national happiness was more important than gross national product. For decades, the country took to measuring a multidimensional gross national happiness index and to spreading the word internationally.

Although the current prime minister of Bhutan has set aside focus on gross national happiness, the idea of alternative metrics for national progress has had global impact. In 2012 at a meeting entitled "Happiness and Well-Being: Defining a New Economic Paradigm," United Nations Secretary-General Ban Ki-moon declared the need for "a new economic paradigm that recognizes the parity between the three pillars of sustainable development. Social, economic, and environmental well-being are indivisible. Together they define gross global happiness" ("Ban: new economic paradigm...," 2012).

In the same year, the United Nations proclaimed March 20 the "International Day of Happiness" to recognize "the relevance of happiness and well-being as universal goals and aspirations in the lives of human beings around the world and the importance of their recognition in public policy objectives."[4] The United Nations' first *World Happiness Report* (Helliwell et al., 2012) provides a detailed analysis of how different countries compare on the happiness scale.

But, of course, measurement is only half the battle. Governments are also engaged in determining how best to use these measures in public policy, and we in the technology field can learn much by observing their various strategies. Encouraging healthy behavior is one approach policymakers have often taken. The UK government's Behavioural Insights Team within the Cabinet Office (referred to as the "Nudge Unit") "applies insights from academic research in behavioural economics and psychology to public policy and services."[5]

Educating the public as to things they can do to improve their own wellbeing is another approach. In 2008, the UK Government Office for Science commissioned a set of evidence-based actions people can take to improve their psychological wellbeing. The idea was to do for mental health what a campaign promoting "five a day" (of fruits and vegetables) had done for physical health. The result was a thorough review (Denham, Beddington, & Cooper, 2008) of extensive research into wellbeing consolidated into "five ways to well-being": "connect," "be active," "take notice," "learn," and "give." Elegant in their simplicity and yet powerful in the strength of the research behind them, the "five ways" could provide a valuable set of pillars for work in positive computing.

Initiatives will continue to emerge and evolve, and they all will face the challenge of deciding how to be guided by new information toward the development of effective and equitable public policy that respects both privacy and autonomy. As with any government decision, the line between where a government should and shouldn't intervene on behalf of national wellbeing will remain a point of ongoing controversy and negotiation. Since large economic gaps within a society produce unhappiness, should we design the tax code to minimize the gap, thus supporting happiness? Should government regulate junk-food advertising targeted to children in order to reduce the harm to wellbeing done by consumption of these foods? If married people tend to be happier according to research, should the government invest in matchmaking? Some decisions will seem easier to rule out than others, but the line will represent a constant point of dialectic—what's most heartening is that we are finally having these discussions.

Even in the United States, where the population is arguably among the most publically averse to tax increases and government intervention, a government's responsibility to promote public happiness has been woven into the very foundations of the country. The Declaration of Independence famously states that we are all divinely entitled to "Life, Liberty, and the *pursuit of Happiness*" and, less famously, "that to secure these rights, Governments are instituted among Men, deriving their just powers from the consent of the governed." B. S. Bernanke, former chairman of the US Federal Reserve gave a commencement address in 2010 that brought these eighteenth-century ideals into modern focus. Entitled "The Economics of Happiness," his address suggested that wealthy countries have the resources to invest in things that contribute to happiness, such as medical care, good nutrition, and sanitation; to maintain a clean environment, and to "provide leisure time and facilities, less physically exhausting and more interesting work, higher education levels, greater ability to travel, and more funding for arts and culture."[6]

Bernanke's commencement advice suggests that organizations focus on supporting conditions for wellbeing by investing in those things that can positively affect it. Richard Thaler and Cass Sunstein (2008) support a "libertarian paternalism" that seeks to help citizens make healthier decisions while preserving autonomy and choice. Wellbeing researcher Nic Marks (who led the creation of the Happiness Index) echoes Bernanke's address in a TED Talk: "Government shouldn't try to make us happy directly, that would be a bit weird. Government should be about making the conditions out of which well-being can emerge."[7] The notions of gently nudging healthier behavior or creating the right conditions for it are two pathways that also will (and already do) guide researchers in positive computing. We look at examples of these pathways in part II of the book.

Where there is much to be learned from the methods and strategies employed in economics and public policy, where keywords such as *nudge*, *expenditure*, and *provision* shape the landscape, a third approach to increasing psychological wellbeing across a population is to be found in education.

Education: Wellbeing as Learnable and Good for Learning

It's estimated that at any given point in time 10 to 20 percent of youth will suffer a mental health problem (O'Connell, Boat, & Warner, 2009). Why do we wait for serious problems to occur before taking action? The lack of preventative and promotional efforts for wellbeing have caused many psychologists and neuroscientists to turn to schooling as an obvious partner in giving people a better and more resilient start in life.

In a recent report of the US National Research Foundation (O'Connell, Boat, & Warner, 2009), numerous leading researchers from across the social sciences call on "the nation—its leaders, its mental health research and service provision agencies, its schools, its primary care medical systems, its community-based organizations, its child welfare and criminal justice systems—to make prevention of mental, emotional, and behavioral disorders and the promotion of mental health of young people a very high priority. By all realistic measures, no such priority exists today."

Although the focus of the report is *prevention*, it also embarks on an analysis of mental health *promotion* through supportive families and schools—the very environments where young people develop the traits that will support their wellbeing and help them manage negative emotion and behavior throughout their lives. The report highlights evidence that the best strategies for preventing cognitive, emotional, and behavioral disorders are early intervention. It calls on the nation first to support those at risk, providing them with the best evidence-based interventions available, and then to promote the development of socioemotional skills in children and young adults more generally.

Just as modern economists and politicians are looking to "measure what matters," educators are interested in "teaching what matters." However, their discussions are understandably dominated by a focus on traditional academic subjects such as literacy, math, and science. Despite significant evidence that youth is the most critical window of opportunity for the development of attributes necessary to a happy stable life, only a small fraction of the efforts of educators, learning scientists, and education policymakers has been directed at developing psychological resources.

In fact, it is far more likely to find school-based peer-reviewed evaluations of wellbeing initiatives in journals such as *Addiction* than in education publications such as the *Journal of Educational Psychology*. The term *wellbeing* for example appears in only twenty articles in the latter, and half of them are from before 1950. The term *mathematics*, on the other hand, appears 795 times. Wellbeing was evidently not the focus of educational psychology in the second half of the twentieth century. Government funding, especially over the past decade, has encouraged a focus on what is collectively referred to as "STEM education," consisting of science, technology, engineering, and math.

Without any doubt, society needs the scientists and engineers who will address the serious challenges of energy, climate change, and future technologies essential to our survival. And, of course, it's also critical that future generations gain the sophisticated understanding in these areas that will allow them to tackle twenty-first-century issues. But it seems that socioemotional skills—which are predictors of success in life—creative problem solving, and better decision making have been undervalued for far too long.

From the policy perspective, not only should adding wellbeing to the curriculum increase national wellbeing measures, it will also attend to other problems governments face, such as crime, poverty, drug abuse, and illness, all of which are frequently born from and exacerbated by ill-being. Intervening only once things are diagnosably bad requires expensive strategies such as long-term treatment and incarceration.

Despite the minimal attention given wellbeing in academic education research, new approaches geared at integrating socioemotional learning into the curriculum are emerging at the level of practice and within policy groups. For example, all children from kindergarten to sixth grade in New South Wales, Australia, follow a curriculum titled "Personal Development, Health, and Physical Education" that includes such modules as "Self and

Relationships," "Own Feelings and Empathy," "Respect and Responsibility," "Dealing with Conflict," and "Diversity." In the United States, the term *social-emotional learning* is used to describe similar curricula that aim to improve relationships and to develop emotional awareness and regulation, self-control, and healthy values. In the National Research Foundation report mentioned earlier, some of these programs were "shown to promote positive youth development while preventing mental health problems as well as substance abuse, violence, and other antisocial behaviour" (O'Connell, Boat, & Warner, 2009). Programs include:

- Inner Kids Program
- Inner Resilience Program
- Mindful Schools Program
- MindUP Program
- Still Quiet Place Program
- Stressed Teens Program
- Wellness Works in Schools Program

A RAND technical report titled *Interventions to Improve Student Mental Health* (Stein et al., 2012) provides an interesting review of the literature on such interventions written for policymakers in California. Prevention and early-intervention initiatives are grouped into those aiming to reduce stigma and discrimination, those on suicide prevention, and those on student mental health.

Others outside universities and governments have also recognized the importance of such research. The largest philanthropic organization in the world, the Bill and Melinda Gates Foundation, recently funded a project led by neuroscientist Richard Davidson aimed at developing mindfulness in children.[8] We discuss this project and other such schools projects in part II.

Of course, the difficulty in attending more seriously to this area of development in schools is compounded by a modern reliance on test scores as measures of student and teacher competence. Yet, remarkably, research shows that, in addition to making happier, safer, more resilient kids, these wellbeing programs also increase their *academic* performance. A recent meta-analysis (Durlak, Weissberg, Dymnicki, Taylor, & Schellinger, 2011) of studies that included more than 200 schools (more than 270,000 students) showed that social and emotional learning programs lead to an impressive 11 percent gain in academic achievement.

This relationship is perhaps not surprising since positive emotions have been linked to better problem solving and enhanced creativity. Don

Norman, among others, has highlighted the importance of designing for emotion (2005), much of which can be applied to the design of learning technologies. In Dorian's book *Interface Design for Learning* (2014), she looks at some of the emotions critical to learning and how design can support these emotions for better learning outcomes. For example, numerous studies have shown that positive emotions increase learning, creativity, problem-solving ability, and big-picture thinking. This work is related to Barbara Fredrickson's research on how positive emotions improve not only life experience, but also efficacy and resilience by increasing awareness, creativity, and exploratory behaviors (we look at the evolutionary importance of positive emotions in chapter 6).

As such, developing wellbeing in the learning environment also benefits from a better understanding of the dynamics of emotion involved in learning experiences as they occur. Educational psychology has tended to focus heavily on the cognitive aspects of learning rather than on the affective phenomena involved. Nevertheless, a number of researchers in the field have worked on certain areas of emotional experience, such as test anxiety, anger, frustration, and self-regulation (in both students and teachers) (Schutz & Pekrun, 2007). Among these emotions, test anxiety has received the most attention (particularly in the past few years in response to the increased reliance on standardized testing measures for evaluation in the United States).

The control-value theory of academic emotions (Pekrun, 2006) provides a way to analyze the antecedents and consequences of what students feel in learning situations. The theory assumes that appraisals of control (what is under a student's control and what is not) and values (how important the task is to a student) are essential to understanding the emotions felt in these activities (e.g., enjoyment, frustration, and boredom related to the learning activity as well as joy, hope, pride, anxiety, hopelessness, shame, and anger related to the outcome of the activity). There are clear overlaps with self-determination theory and its pillars of autonomy and competence. Education and wellbeing research undoubtedly have much to learn from each other, and we anticipate that they will begin to partner more consistently over the coming decade and will in all likelihood increasingly turn to technologists in their search for new tools to support investigation, learning, and wellbeing.

Learning from Learning Technologies

Both of us have spent much of our professional careers developing, evaluating, and researching technologies for learning. One interesting thing about these technologies is that they represent an area in which researchers have begun to combine emotions and technology in at least two ways. Despite the focus on STEM, there are at least two areas in which wellbeing measures have already been incorporated. First, at the convergence of affective computing and learning, we see experimental research in the use of emotionally aware, intelligent tutoring systems—systems that recognize and respond to boredom, confusion, and frustration and that promote engagement and resilience (see Calvo & D'Mello, 2011, 2012). We look at affective computing more closely in the next chapter.

Second, at the intersection of education, technology, and mental health, there exists a body of research on various Internet-based and technology-delivered interventions in schools. For example, one program on alcohol education (Champion, Newton, Barrett, & Teesson, 2012; Newton, Teesson, Vogl, & Andrews, 2010) randomly allocated 764 young teenagers from across ten schools to an Internet course or to a standard face-to-face health class. After 12 months, those who did the online course were found to be more knowledgeable, to consume less alcohol, and to have fewer binge-drinking episodes than those who did the face-to-face class. Although digital programs for personal development and wellbeing in schools have been surprisingly slow to take off (particularly in contrast with how much digital attention has been given to math and literacy), we expect to see growth in this area over the next decade as interest (and funding) in technology intersects with that of wellbeing.

With regard to learning that occurs outside of schools, researchers in the area of interaction design for children (IDC) have been pioneering in their attention to factors of wellbeing. Svetlana Yarosh and her colleagues (Yarosh, Radu, Hunter, & Rosenbaum, 2011) surveyed the papers published in each year of the IDC conference from 2002 to 2010 and sought to understand the type of behaviors and qualities the IDC community tried to promote in children. These behaviors and qualities were broadly grouped into *social interaction and connectedness*, *learning*, *expression*, and *play*, all of which impact on wellbeing. Furthermore, a number of apps for children are intended to promote wellbeing factors specifically. For example, *Focus on the Go!* and *Sesame Street for Military Families* are designed to help military children build resilience skills. *Emotionary* is among a number of apps designed to help kids identify and communicate emotions, and *Positive-Penguins* supports them in challenging their thinking (in the style of cognitive behavioral therapy). For older kids, *Middle School Confidential* is a high-quality app-delivered comic that deals with confidence and bullying issues. Although such trailblazers represent just the beginning, a research field in positive computing will help support further work in this largely untapped area as the field matures.

Education, economics, and policy can help us to measure or influence wellbeing across a group or population, but these fields do less to explain why variations in wellbeing occur in the first place. To understand this higher-level question, we have to look at how various societal and cultural influences shape our wellbeing and our understandings of it.

Social Science: Wellbeing as a Changing Cultural Construct Shaped by Technology

A purely psychological analysis aims to understand wellbeing as an internal positive state we aim to attain. A sociocultural approach, in contrast, places more focus on how the definition of wellbeing changes over time and across cultures. *Technology and Psychological Well-Being* (Amichai-Hamburger, 2009) contains a series of essays exploring the relationships from a social sciences perspective. Work in sociology and media studies is critical to helping us understand how technology has already impacted our wellbeing and why.

George Rodman and Katherine Fry's (2009) historical account highlights wellbeing as a historical and cultural construct. The authors focus on wellbeing's relationship to social connections and discuss how the predominant communication technologies of each culture might have influenced its views on concepts and values such as individuality, society, privacy, and wellbeing. For example, the invention of typography in the 1450s introduced major social and economic changes, and, according to Rodman and Fry's reading of Marshall McLuhan, some of the changes were dehumanizing as they reduced the need for face-to-face interactions, but other changes were liberating as they democratized information and raised the sense of self. Certainly we see similar parallels arising with the spread of modern information communication technologies, some of which are elegantly explored in Richard H. R. Harper's book *Texture: Human Expression in the Age of Communications Overload* (2012).

Social media in particular have made whole new social behaviors possible with both positive and negative consequences to wellbeing. Social media researcher danah boyd, of Harvard and Microsoft Research, points to the life-saving potential provided in parallel with challenges posed by social media with regard to the wellbeing of youth (see her sidebar in this chapter for more detail).

Although economics has helped to uncover the links (or lack thereof) between wealth, technology, and wellbeing, knowing that more wealth or more advanced personal technology hasn't made society much happier doesn't tell us why it hasn't. Perhaps wealth would be a more effective indicator if some other variable were changed. Perhaps technology would have greater positive impact if it were designed differently. The weakness in our current understanding is certainly influenced by the difficulty that exists in isolating variables within such a complex system.

For example, one study that surveyed a cross-section of 22 European countries (Frey, Benesch, & Stutzer, 2007) shows a negative correlation between TV ownership/viewership and wellbeing (more TV time was linked to lower life satisfaction). Another study (Dolan, Metcalfe, Powdthavee, Beale, & Pritchard, 2008) suggests the opposite—that having *TV and computers* improved self-reported wellbeing measures. A third study (Kavetsos & Koutroumpis, 2011) used a cross-sectional database of 29 European countries and found that those who owned a phone, CD player, and computer *and* who had an Internet connection were more likely to report higher subjective wellbeing. In this third study, correlation with TV ownership was statistically insignificant.

There is obviously much we need to learn about which technologies can support wellbeing, when, in what circumstances, in what combinations, and why. Ethnographic and anthropological research, historical and sociological inquiry, along with other methods matured by the social sciences will be essential to moving us forward toward this understanding.

Moving away from the social sciences and toward examples of application, we come to business—an area for which investment in wellbeing poses a clear value proposition.

Business and Organizational Psychology—Wellbeing in the Workplace

In the past 20 years, growing research evidence has shown that happier employees can be more productive, innovative, and empathic with their clients (Goleman, 1998; Linley, Harrington, & Garcea, 2010). This is part of the reason why an increasing number of corporations and nongovernment organizations have begun turning to wellbeing-related training and initiatives in the form of emotional intelligence training, mindfulness and meditation events, as well as changes intended to improve work–life balance, social connectedness, and autonomy.

From the perspective of positive computing, companies are likely to provide many opportunities because they are significant users of online and technology-supported training. In addition, companies now rely on enterprise-level social media and communication platforms, many of which will likely be found to benefit from and feed into wellbeing-informed design.

Human-resource systems are rather sophisticated beasts, recording and analyzing employee performance data and creating strategic maps to guide employee skill sets (knowledge capital) to adapt to changing needs. Yet, there is much room for improvement in how they manage *mental capital*, or an organization's wellbeing skills.

In countries such as the United Kingdom, employers are liable for the consequences of work-related stress. One of the ways of addressing this liability (and helping employees) is to offer counseling services. This approach has persuaded employers to provide training and counseling programs, often delivered by private companies, which arguably has bootstrapped an industry in wellbeing. The University of Sydney, for example, has a Health and Wellbeing Program that includes professional counseling, self-help books and materials, peer-support programs, and a number of health initiatives (e.g., Weight Watchers and a smoke-free environment).

Just as new training requirements arising from health and safety compliance fed the e-learning industry, the same kind of energy toward improved employee wellbeing can now drive an industry of positive computing—for example, in the form of wellbeing-informed redesigns of employee systems. Indeed, there is evidence this is already happening. For example, Crane software by Kanjoya provides managers with information about the mood of their organization and of the groups of people within it. We suspect that we will soon see more examples in this genre—for instance, customer-service software designed to increase empathy between customer and rep or more sophisticated ways of measuring the wellbeing effects of management changes or human-resource programs. As with education, this area presents massive potential for research and practice in positive computing to make significant contributions.

One final way to approach the role of technology in our experience of wellbeing is by viewing technology as part of our environment. Fortunately, the question of how environments impact wellbeing is not new to architects and environmental designers, so as a final multidisciplinary foray we turn to them.

Design and Architecture: Places and Things That Improve Wellbeing

In *The Architecture of Happiness* (2006), Alain de Botton writes about ways in which art and architecture have been used over the ages to influence what we feel, think, and do. A cathedral, for example, in its very proportions, its magnificent statues, and its filtered light invites contemplation and awe in a way that a fast-food restaurant doesn't. In contrast, the restaurant architects favor the emotional effects of modernity, economy, and speed, with visible kitchen staff, pop music, and bright colors that pique the appetite.

Environmental psychology provides a methodological approach to studying the affordances and effects of place (Bell, Greene, Fisher, & Baum, 2005). The field connects research on the physical environment with research on health and wellbeing, investigating ways in which certain designs promote, hinder, or completely rule out certain behaviors. Topics within their sphere range from how the availability of informal spaces can increase a sense of community and reduce criminal behavior to how high ceilings lead to more open-ended thinking (known as the "cathedral effect") and how an office window might increase job satisfaction.

Although environmental psychology focuses on physical rather than digital environments, it suggests what some of the methodological challenges facing positive-computing work will be. Rather than studying perception as a separate phenomenon of the stimuli, environmental psychologists study both as a single entity (similarly, the stimulus–response situation of a website and a user does not depend on just one or the other of them). These ideas can be adapted to digital environments in that interaction phenomena depend on website design, but also on the experiences of the person interacting (i.e., their previous experience, interests, level of education, etc.). In environmental psychology, the environment–perception unit is more than the sum of its parts. This focus means that environmental psychologists, like other psychologists, tend to work in the field rather than in the laboratory. The challenges of field study have led environmental psychologists to use a rich mixture of methods (Bell et al., 2005), some of which are discussed in chapter 6.

Recent work by researchers Pieter Desmet and Anna Pohlmeyer in what they term "positive design" shows how the aim to support psychological wellbeing is playing out in parallel with positive computing within the field of industrial design. In a special issue of the *International Journal of Design*, Desmet, Pohlmeyer, and Jodi Forlizzi (2013) bring together work on design for subjective wellbeing from the perspectives of experience design, business, ethics, and codesign.

. . .

In chapters 2 and 3, we have looked at some of the seminal wellbeing-related work in a diversity of disciplines. The unique perspectives and opportunities that arise from multidisciplinary views will act as rich resources of information as well as fertile areas of future application as work in positive computing moves forward. As always, drawing from and working across multiple disciplines will pose challenges to communication as well as to funding mechanisms, but the rewards are great, and they come in the form of unique solutions, broader perspectives, and greater benefits to users.

In the next chapter, we come back to base and review some of the ways in which researchers in engineering and computer science have already begun to consider wellbeing issues as part of their work in technology.

Expert Perspectives—Multidisciplinary Views

When Worlds Collide: The Power of Cooperation in Wellbeing Science

Jane Burns, Young and Well Cooperative Research Centre

Imagine a research center where young people work with scientists, service providers, technologists, and governments in a quest to find an answer to the question *"Can technology be used to enhance the wellbeing of young people?"* Such a center exists in Australia and is called the Young and Well Cooperative Research Centre (CRC). Funded under the Australian government's CRC program, it unites young people with researchers, practitioners, innovators,

(continued)

and policymakers from more than 70 partner organizations across the not-for-profit, academic, government, and corporate sectors.

The Young and Well CRC fundamentally puts young people in the innovation "hot seat," directly asking how they use technology to enhance their wellbeing and seeking to understand what other technology, new or emerging, they suggest would be beneficial. This model tips on its head the idea that "the answer" lies with the gifted scientist, the behind-the-scenes technologist, or even the creative entrepreneur.

A philosophy embraced by the Young and Well CRC and its partners is that for true innovation to occur, young people must work with scientists, innovators, technologists, young entrepreneurs, and service providers in a world where perspectives collide to spark new ways of thinking. This model of collaboration is fraught with challenges: communicating across multiple organizations, establishing cross-disciplinary teams that bring their own jargon, getting the science right while ensuring that the resources or products for wellbeing aren't compromised, and keeping pace with technologies as the innovation moves faster than the research.

That said, the challenges pale into insignificance when you imagine a world where technologies are embraced in a way that supports the wellbeing of young people. In many ways, our young people are already setting the technologies and wellbeing agenda. You can see it in the way they are building online social networks that are accepting of diversity, are issues based, and, when built appropriately, provide a space that can be safe and supportive and allow a young person to feel valued and connected. Young people are similarly creating digital content, which provides an opportunity to share their thoughts and feelings regardless of gender, race, ability, and literacy levels.

The challenge for each of us—whether a psychiatric epidemiologist, a computer technician, the dean in the Faculty of Health, the CEO of a mental health service, or an educator, psychologist, or social worker—is to embrace the possibilities that collaborative partnerships bring, acknowledge that it is hard, but also accept that we have an opportunity to harness technologies in order to fast-track our approach to wellness.

(continued)

> **Making Sense of Increased Visibility**
>
>
>
> danah boyd, Harvard and Microsoft
>
> Technology allows us to see into the lives of more people today than ever before in history. Because of the public nature of major social media platforms, it's often possible to see traces of strangers' activities, interactions, and interests.
>
> Through Twitter, I can watch a group of Indonesian teens talk about their love of a particular boy band, and on Instagram I can view the photographic trail of a Brazilian twenty-something as she documents her vacation. I can use Google Translate to get a sense of what Chinese youth are talking about on Weibo, and I can traverse profiles of Russian friends and families on VKontakte without even knowing the language.
>
> These images, networks, and status updates never tell the whole story, but they offer glimpses into the lives of people who are quite different than those I meet every day in my personal and professional life. They are not people whom I would encounter by accident, but social media create a digital street for me to stroll down.
>
> I relish the opportunity to learn about the world from varied vantage points, but I also struggle with a slew of ethical challenges that I face as I think about how to make sense of what I see. How do I know that my interpretation of what I see is accurate? In my research on American youth, I regularly found that teens would encode what they wrote. They were happy to

(continued)

> make their content publicly accessible while limiting access to the meaning of what they shared.
>
> I don't always have the contextual information or know the relevant cues to meaningfully interpret the traces that are in front of me. And although I try hard not to be judgmental of what I see, I know plenty of people take what they see out of context. What do I do when what I see is deeply problematic? In my efforts to look into other people's lives, I have seen countless cries for attention, including suicidal proclamations, detailed accounts of self-injury, and lashing out that most likely comes from a place of abuse.
>
> What I have access to are simply traces from people whom I don't know and may not be able to identify even if I tried. Many of the most painful pleas come from people who are anonymous online. Are they really experiencing what they state? Is there anyone watching? Are they going to be able to get help?
>
> The visibility of people's lives through social media is both a blessing and a curse. On one hand, seeing diverse experiences offers valuable insight, and the potential to connect across traditional barriers is greatly increased. On the other, many traces reveal that there are people who are seeking love, support, and attention but aren't finding what they need. How can we leverage visibility to enable eyes on the digital street? How can we use what we see to increase people's access to support and services and otherwise increase people's wellbeing?
>
> Rather than looking at social media with disdain, it's important to start by opening our eyes. I recommend that you turn to your favorite platform, whether it's Twitter or Tumblr, and spend time looking at the traces left by strangers. Rather than being horrified or disgusted, ask yourself a simple question: What is it about this person's life that makes posting this message completely sensible? Step back and appreciate difference. And when it's clear that someone is hurting, ask another question: What can be done to help this person or other people like him or her feel stronger, happier, and more supported? The more we individually do to make people's lives better, the more society wins.

Notes

1. From the website beyond-gdp.eu.

2. See happyplanetindex.org.

3. See well-beingindex.com.

4. From un.org/en/events/happinessday/.

5. From the Behavior Insights Team website at gov.uk/government/organisations/behavioral-insights-team.

6. See Bernanke's address at federalreserve.gov/newsevents/speech/bernanke 20100508a.htm.

7. See "Nic Marks: The Happy Planet Index," at ted.com/talks/nic_marks_the_happy_planet_index.htm.

8. For this project, see news.wisc.edu/releases/17368.

References

Amichai-Hamburger, Y. (Ed.). (2009). *Technology and psychological well-being.* Cambridge, UK: Cambridge University Press.

"Ban: New economic paradigm needed, including social and environmental progress." (2012). UN News Centre. April 2. Retrieved March 2014, from http://www.un.org/apps/news/story.asp?NewsID=41685#.UzOETCTvWkB.

Bell, P. A., Greene, T., Fisher, J. D., & Baum, A. (Eds.). (2005). *Environmental psychology.* Vol. 4. 5th ed. Fort Worth, TX: Routledge.

Bruni, L., & Porta, P. L. (Eds.). (2007). *Handbook on the economics of happiness.* Malden, MA: Edward Elgar.

Calvo, R. A., & D'Mello, S. (Eds.) (2011) *New perspectives on affect and learning technologies.* New York: Springer.

Calvo, R. A., & D'Mello, S. (2012). Frontiers of affect-aware learning technologies. *Intelligent Systems, IEEE, 27*(6), 86–89.

Champion, K. E., Newton, N. C., Barrett, E. L., & Teesson, M. (2012). A systematic review of school-based alcohol and other drug prevention programs facilitated by computers or the Internet. *Drug and Alcohol Review, 2013*(32), 115–123.

De Botton, A. (2006). *The architecture of happiness.* New York: Pantheon.

Denham, J., Beddington, J., & Cooper, C. (2008). *Mental capital and well-being project.* London: UK Office for Science.

Desmet, P. M. A., Pohlmeyer, A. E., & Forlizzi, J. (2013). Special issue editorial: Design for subjective well-being. *International Journal of Design, 7*(3). Retrieved from http://www.ijdesign.org/ojs/index.php/IJDesign/article/view/1676/594.

Diener, E., & Seligman, M. E. P. (2004). Beyond money. *Psychological Science in the Public Interest, 5*(1), 1–31.

Dolan, P., Metcalfe, R., Powdthavee, N., Beale, A., & Pritchard, D. (2008). *Innovation and well-being*. Innovation index working paper. London: Nesta.

Durlak, J. A., Weissberg, R. P., Dymnicki, A. B., Taylor, R. D., & Schellinger, K. B. (2011). The impact of enhancing students' social and emotional learning: A meta-analysis of school-based universal interventions. *Child Development, 82*(1), 405–432.

Easterlin, R. A. (1974). Does rapid economic growth improve the human lot? Some empirical evidence. In P. A. David & M. W. Reder (Eds.), *Nations and households in economic growth: Essays in honor of Moses Abramovitz* (vol. 8, pp. 88–125). New York: Academic Press.

Frey, B. S., Benesch, C., & Stutzer, A. (2007). Does watching TV make us happy? *Journal of Economic Psychology, 28*(3), 283–313.

Golder, S. A., & Macy, M. W. (2011). Diurnal and seasonal mood vary with work, sleep, and day length across diverse cultures. *Science, 333*(6051), 1878–1881.

Goleman, D. (1998). *Working with emotional intelligence*. New York: Bantam.

Harper, R. H. R. (2012). *Texture: Human expression in the age of communications overload* (p. 320). Cambridge, MA: MIT Press.

Helliwell, J., Layard, R., & Sachs, J. (2012). *World happiness report*. New York: Earth Institute.

Kahneman, D., Diener, E., & Schwarz, N. (Eds.). (1999). *Well-being: The foundations of hedonic psychology*. New York: Russell Sage Foundation.

Kavetsos, G., & Koutroumpis, P. (2011). Technological affluence and subjective well-being. *Journal of Economic Psychology, 32*(5), 742–753.

Linley, A., Harrington, S., & Garcea, N. (Eds.). (2010). *Oxford handbook of positive psychology and work*. New York: Oxford University Press.

Mitchell, L., Harris, K. D., Frank, M. R., Dodds, P. S., & Danforth, C. M. (2013). The geography of happiness: Connecting Twitter sentiment and expression, demographics, and objective characteristics of place. *PLoS ONE, 8*(5), 15.

Newton, N. C., Teesson, M., Vogl, L. E., & Andrews, G. (2010). Internet-based prevention for alcohol and cannabis use: Final results of the Climate Schools course. *Addiction, 105*(4), 749–759.

Norman, D. A. (2005). *Emotional design: Why we love (or hate) everyday things.* New York: Basic Books.

O'Connell, M. E., Boat, T., & Warner, K. (2009). *Preventing mental, emotional, and behavioral disorders among young people: Progress and possibilities.* Washington, DC: National Academies Press.

Pekrun, R. (2006). The control-value theory of achievement emotions: Assumptions, corollaries, and implications for educational research and practice. *Educational Psychology Review, 18,* 315–341.

Peters, D. (2014). *Interface design for learning: Design strategies for learning experiences.* San Francisco: New Riders.

Rodman, G., & Fry, K. G. (2009). Communication technology and psychological well-being: Yin, yang, and the golden mean of media effects. In Y. Amichai-Hamburger (Ed.), *Technology and psychological well-being* (pp. 9–33). Cambridge, UK: Cambridge University Press.

Ryan, R. M., & Deci, E. L. (2001). On happiness and human potentials: A review of research on hedonic and eudaimonic well-being. *Annual Review of Psychology, 52,* 141–166.

Schutz, P. A., & Pekrun, R. (2007). *Emotion in education.* Salt Lake City: Academic Press.

Stein, B. D., Sontag-Padilla, L., Chan Osilla, K., Woodbridge, M. W., Kase, C. A., Jaycox, L. H., … Golan, S. (2012). *Interventions to improve student mental health: A literature review to guide evaluation of California's mental health prevention and early intervention initiative.* Santa Monica, CA: RAND.

Thaler, R. H., & Sunstein, C. R. (2008). *Nudge: Improving decisions about health, wealth, and happiness.* New Haven, CT: Yale University Press.

Yarosh, S., Radu, I., Hunter, S., & Rosenbaum, E. (2011). Examining values: An analysis of nine years of IDC research. In *10th International Conference on Interaction Design and Children. IDC 2011* (pp. 136–144). Ann Arbor: University of Michigan.

4 Wellbeing in Technology Research

Computers should be able to do *X*. Current techniques only do *X-I*. We contribute a technique that does *I*.

It's not exactly an inspiring narrative, but this humble argument has nevertheless fueled incremental technological progress over the past century and ushered us through three generations of computing and into the Internet of Things we find ourselves moving today. As devices get embedded into the fabric of our lives and become inextricable parts of the experiences that shape us, their inevitable impact on our wellbeing grows ever greater. Yet engineering hangs onto technology-focused approaches. Sometimes humans are included in the equation, although mainly as comparison points:

Computers should be able to do *X*. Humans do *X* very well. Current techniques do only *X-I*. We contribute a technique that does *I* by emulating the way a human does it.

I (Rafael), for one, have used both of these arguments in my work (just replace *X* with "recognize emotions" or "help students" and *I* with "use language" or "give feedback"). Yet any system created for human use will have some effect on human psychological wellbeing, however profound or negligible. In order to be able to take this effect into account in our design, we can start by looking at how current technologies already impact wellbeing and at how current research areas can or already do contribute to our understanding in this area.

Ubiquitous Computing: Opportunities and Challenges for Wellbeing

Two decades ago in a seminal paper, Mark Weiser (1991) coined the term *ubiquitous computing*, now affectionately abbreviated to "ubicomp." He simultaneously predicted many of the technology developments that would come to characterize modern digital experience. He was dissatisfied with the contemporary personal-computing model and argued that computing devices could fulfill their potential most effectively through seamless integration into the world: "My colleagues and I at the Xerox Palo Alto Research Center think that the idea of a 'Personal' Computer itself is misplaced and that the vision of laptop machines, dynabooks and 'knowledge navigators' is only a transitional step towards achieving the real potential of information technology."

Weiser's article both reflected and shaped an important era of technology work. The ubicomp vision has nurtured myriad academic projects, journals, and conferences. The technical challenges of building these systems are formidable and have been reviewed with regularity (Abowd & Mynatt, 2000; Estrin, Govindan, Heidemann, & Kumar, 1999; Yick, Mukherjee, & Ghosal, 2008). The future, thanks to ubicomp, promises to bring us constant access to information and computational capabilities, together with new ways of interacting with them. The challenge now is figuring out how best to recruit these new capabilities to best serve human needs.

Some would undoubtedly argue that because many of these technologies are designed to support safety, productivity, and enjoyment in ways that are more sophisticated, more readily available, and more personalized than ever before, they will naturally lead to happier lives. Perhaps—but as we have argued previously, relying on assumptions or using proxies such as productivity or personalization for wellbeing is a weak approach to a future of human-centered engineering. We now have the tools and theory it takes to be more precise in our evaluations and to inform our design more consciously; in other words, we can do better than just relying on assumptions in ensuring that technology fosters flourishing.

Two decades have gone by since the ubicomp vision entered the world, and we now have what it takes to move genuine human-centeredness to the next level. One area of development has already begun to explore the potential for data collection, evaluation, and reflection to support personal growth, and it's known as "personal informatics."

Personal Informatics—New Tools for an Old Quest

Sometime around 400 BCE, a passing Greek paused at the Temple of Apollo long enough to inscribe these timeless words into its stone: "Know Thyself." This aphorism was thereafter interpreted in at least three different ways by the ancient Greeks to follow. Socrates invokes the inscription to explain why the pursuit of self-knowledge is more important than any other intellectual pursuit: "I am not yet able, as the Delphic inscription has it, to know myself; so it seems to me ridiculous, when I do not yet know that, to investigate irrelevant things."[1] In a second interpretation, the maxim is better translated as "Know thy place" or "Who do you think you are?" and is used as a means of restoring modesty to those who need it. Finally, it was used as a reminder that we should aim to know truth *directly* rather than blindly relying on the status quo. "Pay no attention to the opinion of the multitude and revalue not the truth but the accepted custom."[2] There is certainly something to be learned from each of these interpretations, but it is the first that seems to have most powerfully sparked the imaginations of technologists over the past few years.

Life logging, the now familiar digital recording of life events, actually began in the 1980s. Steve Mann at the University of Toronto strapped laboratory equipment to his body and proceeded to diligently record physiological and video data of all his daily activities. In 1994, he went public and started webcasting video of his everyday life, and, like a good convenience store, he remained open seven days a week, 24 hours a day, inviting the world to drop in. Despite Mann's pioneering work in wearable computing, life logging didn't see mainstream uptake until the turn of the century, when the necessary equipment became more affordable and more acceptable to be seen in.

The proliferation of mobile digital devices have seen life-logging tools break out of research labs and move stylishly into the jogging hands of the masses. The use of blogs, microblogs, status updates, cameras, GPS, and other low-cost tools for recording and sharing personal events has become increasingly commonplace.[3] In 2007, Gary Wolf and Kevin Kelly of *Wired* magazine recognized this growing interest in tools that support self-knowledge through "self-tracking" and helped consolidate that interest into a thriving movement that they called the "quantified self" (recruiting Socrates's interpretation of the Delphic inscription as a kind of motto).

Several years on, hundreds of dedicated consumer products allow you to record your movements, sleep patterns, eating habits, and other behaviors. Some quantified-self applications are built as extensions to existing technologies leveraging built-in features of standard devices such as accelerometers, GPS, and gyroscopes, whereas others are separate custom tools. Large sports brands such as Nike and Adidas produce sensors and software to monitor and collect physical activity data. The Fitbit, one of the first sensors to be widely popular, uses a low-cost accelerometer to record your movement walking, running, and sleeping.

In addition, standardized formats are allowing the data produced by these devices to be effectively merged. For example, heart rate and blood pressure data from sensor watches can be synchronized with data from gym machines and wireless scales and then uploaded to one of the many websites available (e.g., Movescount) for people to share their athletic lives, struggles, and success stories.

We have even brought man's best friend into the mix. Modern dog-collar technology allows owners to track Fido's activity level, geographic location, and even happiness throughout the day. If he's at home and you're at work, you can check his activity monitor remotely (opening up whole new opportunities for procrastination). Who knows, maybe in the future we'll be able to sync pet emotional data with our own and finally learn how to be as consistently jolly as dogs are.

But how do all these data get turned into useful information? Log data generally get transferred to a computer, where they can be visualized, analyzed, and shared with others. Companies typically store data for free so that they can mine, process, and transform the data into information that is useful both to users and to company profits.

"Self-trackers" describe experiences in which tracking consumption, medication, and activity data allows them to unearth the causes of health problems. Ian Li, Anind Dey, and Jodi Forlizzi (2010) surveyed 68 self-trackers (and interviewed 11 of those) in order to improve our understanding of what motivates and deters tracking. Most of the tracking activities reviewed had to do with personal finances, energy use, exercise, or work, and their primary motivations to engage included general curiosity and social influence. Findings led the authors to propose five stages in the way people engage with personal informatics technology:

• **Preparation** includes the point at which the decision to track is made along with any associated activities (e.g., deciding on what tool to use).
• **Collection** is when the user records data points that can occur on various time scales, such as hourly or yearly (e.g., food at each mealtime or books you have read over months). This stage is characterized by the technical challenges associated with gathering large sums of data from the user with minimal effort and intrusion.
• **Integration** includes the processing strategies that allow a person or the system to build a meaningful synthesis or visualization from the various data sources.

- **Reflection** occurs when the user reflects on his or her behavior based on the data, and this reflection can happen in real time as in "How many steps have I just walked?" or after the fact as in "How many hours a day have I been walking this month?"
- **Action** is the phase most closely related to the challenges of positive computing. It is here where Li, Dey, and Forlizzi ask, "What are the effects of personal informatics on daily life?" and list aspects such as "trust in the system, motivation, better decision making, loss of control, etc.," some of which are not directly related to wellbeing.

In part II, we look more deeply at the reflective thinking that personal-informatics technologies can support as well as at how reflection can, in the right circumstances, lead to increased wellbeing.

The quantified-self movement, (a.k.a. "personal Informatics," "self-surveillance," "self-tracking," or "personal analytics") has been wildly successful on many fronts. Thousands participate in the online communities, meeting as part of groups around the world, and the movement has received extensive mainstream press coverage. It is driving a significant amount of innovation, academic research, commercial enterprise, and ideally, positive personal change.

But the full story is only beginning to take shape. The workshop on personal informatics held at the Association of Computing Machinery's Computer–Human Interaction conference has sought to develop the dialogue between those in "design, ubiquitous computing, persuasive technology and information visualization" (Li, Dey, Forlizzi, Höök, & Medynskiy, 2011) who are involved in personal informatics. Psychologists are conspicuously missing from the list, despite the fact that psychological impact and issues such as self-awareness, motivation, self-esteem, balance, frustration, pride, self-criticism, ironic processes, and wellbeing are at the core of these digital experiences. Research such as the study by Li, Dey, and Forlizzi has helped illuminate many of the technical obstacles people face at each stage of self-tracking, but research on psychological barriers and variations to experience will be critical to future work. Yvonne Rogers at University College London discusses this point further (see her sidebar in this chapter).

Deborah Lupton (2012), a sociologist at the University of Sydney, has explored how digital technologies affect the people who use them, including their experiences of embodiment, selfhood, and social relationships. Lupton describes self-tracking using "m-health" devices as a conceptual shift in health promotion. On the one hand, digital self-tracking brings a great deal of technology into an area that has been largely low tech and focused on prevention campaigns. These technologies (i.e., social networks, mobile phones, tracking devices) now allow messages to be better personalized.

On the other hand, Lupton cautions that the design of these technologies requires multiple perspectives—technical professionals will focus on the technical problems of a device, and health professionals will look at how these tools can be used efficiently for the treatment and management of medical conditions. However, personal values and moral and ethical concerns must also be addressed. Lupton's (2012) sociological work, in her own words, deals with

how these technologies may operate to construct various forms of subjectivities and embodiments, including identifying the kinds of assumptions that are made about the target of these technologies and what the moral and ethical ramifications of using them may be. Moral implications include the kinds of meanings and the representation of the ideal subject that are related to the use of these technologies in the interests of promoting health. Ethical issues include questioning the extent to which health promotion practice should intrude into their targeted populations' private lives and what kinds of messages and practices they employ when using digital surveillance devices.

Clearly, if we want to see the field mature, we need to share it with those in the social sciences who have academic knowledge in the kinds of human experience we hope to support. For some initial examples of collaboration models for working in multidisciplinary ways as technologists, we might look to affective computing, a field that has studied emotions as well as how to detect, influence, and model them in the context of HCI.

Affective Computing—Technology and Emotions

It wasn't until the early 1990s that computer scientists begun to take emotions more seriously when a small number of researchers started developing computer systems that could detect human emotions. Rosalind Picard at the MIT Media Lab crystallized an emerging interest in *affect* in her seminal book *Affective Computing* (1997). She described three types of affective computing applications:

1. **Affect detection**, in which the computer uses video, microphones, physiological sensors, posture sensors, and other sensing devices to recognize emotions (via facial expressions, voice modulation, posture, etc.). These data are used to train a classifier that maps patterns into emotional

dimensions or labels—for example, the so-called six basic emotions (Ekman, 1992): anger, surprise, happiness, disgust, sadness, and fear.

2. **Affect expression**, in which software agents (e.g., avatars in virtual-reality environments) are able to express emotions. By doing so, users can establish closer relationships and receive more natural feedback.

3. **Emotional computers**, in which a new kind of computer capable of feeling (mechanically embodying) and expressing emotions (albeit machine versions of them) would be developed.

Most affect-detection systems (Calvo & D'Mello, 2010) extract key features from recorded data, then build computational models that map those features to actual emotion labels (or to coordinates if using a dimensional model). In order to generate those models, a classifier is built using machine-learning techniques. The classifier is trained on data collected through a number of possible techniques. Techniques that produce more "ecologically valid" data (i.e., closer to real life) are harder to obtain in laboratory conditions. The most commonly used techniques either interrupt participants in whatever they are doing to ask them how they feel (i.e., to label the training data, often using a standard form or diagram) or ask them once they have finished.

The area has matured into a thriving field with a dedicated journal, a regular conference, and threads within leading HCI journals and conferences.[4] Several reviews (Calvo & D'Mello, 2010) and the *Oxford Handbook on Affective Computing* (Calvo, D'Mello, Gratch, & Kappas, 2014) are further indications of the field's advancement. What is possibly more important for positive computing is that the field's explicit focus on emotions has facilitated strong relationships with other research communities in psychology, psychiatry, education, and neuroscience.

There are a number of examples of affective-computing projects that venture beyond improving HCI. In the next section, we describe three cases in which affective-computing technologies (specifically those for detection) have been used to support psychological wellbeing, the first in the area of attentive technologies, the second in the context of learning, and the third in mental health.

Affect and Attentive Interfaces—When Systems Are Considerate of Your Mental State

Concerned by the capacity for new technologies to produce cognitive overload, a group of researchers turned to the development of what they call "attentive user interfaces" (Vertegaal, 2003). By tracking the user's gaze, an interface can adapt, highlighting urgent issues while backgrounding those of less importance. The significance of such systems is most obvious in technology-rich environments where managing cognitive load is crucial, such as emergency and air-traffic-control rooms. These systems can also be used to study different psychological phenomena—for example, attention, engagement, and mind-wandering—that, as we will see in chapter 8, are related to wellbeing.

Affective Computing for Reflection
One particularly fruitful point of intersection between affect and reflection is in the area of writing. According to research, it's extremely difficult to be writing one thing and thinking another (though we certainly try). This makes writing a uniquely interesting proposition for studying what someone is thinking at a given moment. It's also much easier to analyze writing in the twenty-first century than it ever has been because most writing and the analysis of it are now done on digital devices connected to the Internet. These same devices enable writers to interact with content and other people in completely new ways as part of the writing process. At the convergence of these features, the door is wide open to whole new data-collection channels and new opportunities for understanding ourselves, how we write, and how we think.

It is within this context that my students and I (Rafael) have been working on what we call "Data-Rich Writing Studios" (Calvo, 2014). These systems allow for a combination of work on software architectures, multimodal sensing and fusion, and machine learning for data processing to study writing phenomena in a holistic way, taking into account the writer's physical and social surroundings as well as her cognitive processes and affective states. This information can then be returned to the user in the form of various types of feedback for reflection and learning.

Take, for example, a tool called Glosser, which is a web-based framework for providing automated feedback on writing (Calvo & Ellis, 2010; Villalón, Kearney, Calvo, & Reimann, 2008). Glosser analyzes cognitive aspects of writing, including the development of argument, structure, and topic coverage in order to produce a wide range of feedback. The feedback provided can be on surface or content features, on the writing product (the final document), or on the process itself and is presented as text and visualizations. A quick processing of your essay might reveal at a glance that you haven't covered all topics adequately, that there is a lack of flow from one paragraph to the next, or that the three main arguments you have made lead nicely to the conclusion.

Although this area of research has been focused on education rather than on wellbeing, a number of instructive similarities can be drawn between the two. For example, subjective areas (such as writing and psychological wellbeing) don't involve simple universal right or wrong answers. Therefore, a key design principle has been to provide *reflective* rather than *directive* feedback ("consider this" rather than "do this"). Furthermore, feedback designed to trigger reflection that is based on written text lends itself to use in cybertherapy (an area that already relies heavily on reflective writing in journals). Add to this an ability to detect emotional states from text mining and facial expressions, and whole new opportunities for mental health promotion emerge.

Affect and Technology for Mental Health

The use of writing in therapy is based on research that suggests that writing about thoughts and feelings associated with an experience is beneficial to some individuals. J. W. Pennebaker (1997, 2004) at the University of Texas, Austin, has performed many of these studies. The precise size of these activities' effect is still debated, but it is generally agreed to be positive for physical and mental health.

One of the difficulties is that there are various ways of structuring these writing activities that are helpful to different people. For example, in one study, Laura King and Kathi Miner (2000) found that writing about the positive benefits of an upsetting past experience was beneficial to health. The suggestion is that such structure may enhance self-regulation skills and foster a sense of self-efficacy. The evidence provided by these studies can inform activities and tools for the psychological development of people within clinical scenarios but can also be extended to the design of tools for everyday life.

Another way in which affective computing has been used in psychotherapy is the enhancement of virtual-reality environments. Timothy Bickmore (see his sidebar in chapter 10) and Giuseppe Riva are two researchers spearheading work in this area. Riva (2005) has argued that using virtual-reality exposure therapy for the treatment of anxiety disorders (e.g., fear of heights and speaking in public) is safer, less embarrassing, and less costly than reproducing real-world situations. In this technique, a client is confronted with the stimuli in a way that allows anxiety to attenuate over time. In real life, every time a client avoids situations that cause the anxiety, the phobia is reinforced. In treatment, each successive exposure to the stimuli reduces anxiety through habituation.

Researchers are also exploring how the ever-present mobile phone might also be used for interventions of various kinds for the promotion of wellbeing. For example, one application sends daily motivational text messages based on a number of different psychoeducation campaigns (e.g., stress, random acts of kindness, etc.); another provides mindfulness exercises; and yet another aims to increase sociopolitical participation. Riva and his colleagues have used mobile phones to reduce student stress during exam periods (Preziosa, Grassi, Gaggioli, & Riva, 2009) and commuter stress (Grassi, Gaggioli, & Riva, 2009). More studies that use mobile devices in various ways to reduce stress and improve wellness and wellbeing are regularly published in journals such as *Cyberpsychology, Behavior, and Social Networking*, the *Journal of Medical Internet Research*, and *The Lancet*. Some of these studies involve campaigns to change behavior and offer examples from that increasingly influential area of research and practice known as behavior change technology.

Behavior Change Technology

In the last decade, swelling interest in behavioral economics has encouraged growth in the design of technology that persuades, influences, or helps people to change their behavior. It is self-evident that work in the area of behavior change is highly relevant to positive computing since some improvements to wellbeing will involve supporting this change. Behavior change technology (BCT) (also referred to as "persuasive technology," "captology," and "behavior design") in no way prescribes wellbeing intentions and is applied broadly from advertising to business and politics. However, a large percentage of research in BCT aims to improve wellbeing (often physical aspects) and sustainability.

Researchers focusing on using technology for wellbeing-related behavior change draw on various behavior theories and models including nudge theory, the Transtheoretical Model, the Theory of Planned Behavior, and SDT among others (see Hekler, Klasnja, Froehlich, & Buman, [2013] for a review of behavior theories and models in HCI).

Some work in this area, especially within the category of persuasive technology, can slide into a rhetoric of designer-control and is happily applied by business to increase profitable behavior. Thus, implications for unethical use follow closely behind any discussion of these methods. As such, researchers are working to outline ethical guidelines (Atkinson, 2006, Spahn, 2011, Davis 2010). For positive computing, part of addressing misuse will emerge from the field's definitive aim to support psychological wellbeing, and the imperative to provide evidence (via established multi-dimensional measures) for that claim in practice.

In addition to ethical concerns and issues of user autonomy, we will need to join those researchers in BCT who are challenging the quick-fix thinking that neglects complex, difficult, and long-term change. Martin A. Siegel and Jordan Beck (2014) discuss behavior change technology for quality-of-life improvement advocating for greater acknowledgement that much change is slow and occurs within systems that are complex. They provide the groundwork for an ongoing theory and practice of interaction design for slow changes that they define as "attitudinal and behavioral changes that are difficult to initiate and sustain," bringing to light ethical dilemmas, impacts of timescale, and the value of systems thinking inherent to slow change problems.

Values-Sensitive Design: Acknowledging the Role of Values

Technology may be able to influence people, but should it? If it's going to be impacting people's lives, shouldn't they be involved in deciding how it will do so? Whose values are embedded into every design? Batya Friedman and Peter Kahn (1992) brought long-awaited attention to an elephant that had found a home in the computer science room: values. What is responsible computing, and how can it be promoted among technology designers?

One argument for bringing attention to values is that although "humans are capable of being moral agents and computational systems are not" (Friedman & Kahn, 1992), computers can distort moral agency in two ways: first, when a human's sense of moral agency is undermined by the computer system (as when a human's role becomes secondary and causes him to loose connection to the purpose or meaning of his actions); second, when the computer system projects "intentions, desires and volitions," so that the human makes the computer responsible for his actions.

For more than two decades, the VSD community has asked what computers *ought to* do (rather than what they *can do*) as part of HCI (Friedman, 1996, 1997; Sellen, Rogers, Harper, & Rodden, 2009; Yarosh, Radu, Hunter, & Rosenbaum, 2011). VSD reminds those developing technology that it is impossible to do so without making decisions based on implicit and explicit values and that the values of both designers and users should be accounted for.

VSD literature has been shaped largely by the moral domain of social knowledge (Friedman, 1997). This moral domain considers views and values on justice, fairness, and human welfare. Value-sensitive designers address conflicts between the individual and society, and they investigate values such as privacy, trust, ownership, and health.

Friedman and colleagues (Friedman, Kahn, & Borning, 2006) suggest a series of guidelines for designing computer systems:

- Start with what is most important to you: value or technology or context.
- Identify those who will use the system (direct stakeholders) and others who will be affected by it (indirect stakeholders).
- Identify how your system will benefit or harm each stakeholder.
- Map each benefit and harm onto a list of values.
- Learn about your key values. (VSD recommends reading the philosophical-ontological literature that may provide a definition and ways of assessing it empirically.)
- Identify conflicting values—for example, trust versus security, environmental sustainability versus economic development, privacy versus security, and hierarchical control versus democratization.
- Incorporate values into the organizational structure so that your company can support such initiatives.

These authors also list a set of human values that can be taken into account in system design, including ownership and property, privacy, trust, usability, human welfare, and autonomy. Friedman (1996) has also paid significant attention to autonomy, a topic central to some models of wellbeing that we come back to in part II.

As one might expect, many of the values held by communities are strongly related to wellbeing; however, a link to psychological wellbeing is not required for values to be held. On the flip side, some aspects of psychological wellbeing are not explicitly included in societal or individual values (both mindfulness and resilience, for example, are causally related to psychological wellbeing but are not necessarily conscious core values in most cultures or within common user groups such as teenagers and company employees). The starting point for VSD researchers is moral philosophy and ethics (Friedman et al., 2006), and the sociocultural approach used in VSD can effectively function independently of psychological drivers. For example, Roberto Verganti's (2008) design-driven innovation is based on the capacity to "understand, anticipate and influence emergence of new product meanings." These sociocultural perspectives as well as others, including user-centered design, will continue to provide valuable pieces of the wellbeing puzzle. Clearly there is considerable promise in a future industry that employs both VSD and positive-computing approaches from their complimentary perspectives toward human-centeredness.

Wellbeing in Technology Research

Innovations and inspiration for work in wellness and wellbeing emerge daily from across the many varied technology fields. This chapter was by necessity more of a sample plate than a comprehensive handbook on all that might inform positive-computing work, but to keep informed, you can drop by the website positivecomputing.org to read about or share new examples as they emerge. Now, with a solid grasp of the foundational literature under our belts, in the next chapter we get to the nitty-gritty and look at how all the psychology research on wellbeing might be operationalized into a research and practice framework for future work in positive computing.

Expert Perspective—Technology Research and Wellbeing

Is a Diet of Data Healthy?

Yvonne Rogers, University College London

Digital technology has pervaded all aspects of our lives. Not only does it enable us to access and interact with information and each other, but it can also sense, monitor, inform, and influence human behavior in unprecedented ways. New movements are being established (e.g., the quantified-self and big-data movements) that are rethinking how burgeoning data can be effectively collected, analyzed, and represented to enable the general public, organizations, and government to record, track, and compare using analytic tools, interactive visualizations, and crowdsourcing. But are data availability and

(continued)

accessibility enough? Will finding out more about ourselves through a diet of data lead to more or less happiness?

On the one hand, there is much excitement about how best to exploit, represent, and act upon the explosion of data to improve the quality of life. Some people have started to record their activity patterns (e.g., hours slept, cups of coffee consumed) and to use these notes to improve their behaviors, such as going to bed at different times or drinking less coffee at home. On the other hand, there is the danger that having too much pervasive data can result in information overload, an invasion of privacy, and self-obsessiveness. How can we ensure that new forms of data will be harnessed to good effect while ensuring people remain safe and comfortable? What techniques are available that can transform data so as to empower people and change their behavior in ways that are acceptable and desirable to them?

The opportunities for sensed, streamed, and tracked data for a modern society are immense. The goal of much big-data system development, however, has been to help businesses become more competitive. There is much talk about how data can be turned into "actionable information," increasing profit margins and finding new revenue. Hence, the focus to date has been on improving business models rather than on enhancing the quality of life. In contrast, there has been a paucity of research into how new insights about big data can feed into enhancing people's working and everyday lives. What is needed is a better understanding of how to analyze patterns of user data, from the user perspective and for the user, where users themselves can interact with new information visualizations about their own behavior and decide how to change it. For example, is it possible for people to change their life–work balance through developing different activity patterns when interacting with digital technologies and through developing ways of harnessing the new forms of data to suit their needs?

A challenge for HCI, therefore, is to consider how best to optimize happiness, creativity, and productivity through tapping into and representing the new streams of data in user-meaningful ways. A new approach to technology-based behavior change is needed that focuses on how data in their various forms can be analyzed, modeled, and represented to optimize human life. Ideally, this approach will draw from a number of relevant disciplines—namely, psychology, economics, design, ethics, and computer science—in order to develop new insights, tools, and design guidance that people can act upon to change their lives for the better.

Notes

1. Socrates says this in Plato's *Phaedrus*, 229e.

2. Found in the entry "gamma, 334" of the Suda translations of the Stoa Consortium and explained in *Wikipedia*. The Suda is a massive tenth-century Byzantine Greek historical encyclopedia of the ancient Mediterranean world. See /stoa.org/sol.

3. For one of the earliest examples of life logging, see Buster Benson's *Why I Track* video presentation at vimeo.com/54881153.

4. The journal is *IEEE Transactions on Affective Computing*, and the conference is the International Conference on Affective Computing and Intelligent Interaction.

References

Abowd, G. D., & Mynatt, E. D. (2000). Charting past, present, and future research in ubiquitous computing. *ACM Transactions on Computer-Human Interaction, 7*(1), 29–58.

Atkinson, B. M. C. (2006). Captology: A critical review. In IJsselsteijn, W., de Kort, Y., Midden, C., Eggen, B., & van den Hoven, E. (Eds.) Persuasive technology. Proceedings of the 1st international conference on persuasive technology, pp 171–182. Berlin Heidelberg: Springer.

Calvo, R. A. (2014). Affect-aware reflective writing studios. In R. A. Calvo, S. K. D'Mello, J. Gratch, & A. Kappas (Eds.). *Oxford handbook of affective computing*. New York: Oxford University Press.

Calvo, R. A., & D'Mello, S. K. (2010). Affect detection: An interdisciplinary review of models, methods, and their applications. *IEEE Transactions on Affective Computing, 1*(1), 18–37.

Calvo, R. A., D'Mello, S. K., Gratch, J., & Kappas, A. (Eds.). (2014). *Oxford handbook of affective computing*. New York: Oxford University Press.

Calvo, R. A., & Ellis, R. A. (2010). Student conceptions of tutor and automated feedback in professional writing. *Journal of Engineering Education, 99*(4), 427–438.

Davis, J. (2010). Generating directions for persuasive technology design with the inspiration card workshop. In T. Ploug, P. Hasle, & H. Oinas-Kukkonen (Eds.), Proceedings of the 5th International Conference on Persuasive technology, PERSUASIVE 2010 (pp. 262–273). Berlin: Springer.

Ekman, P. (1992). An argument for basic emotions. *Cognition and Emotion, 6*(3–4), 169–200.

Estrin, D., Govindan, R., Heidemann, J., & Kumar, S. (1999). Next century challenges: Scalable coordination in sensor networks. In *Proceedings of the 5th annual ACM/IEEE International Conference on Mobile Computing and Networking* (pp. 263–270). New York: ACM.

Friedman, B. (1996). Value-sensitive design. *Interaction, 3*(6), 16–23.

Friedman, B. (Ed.). (1997). *Human values and the design of computer technology* (p. 320). Cambridge, UK: Cambridge University Press.

Friedman, B., & Kahn, P. H. (1992). Human agency and responsible computing: Implications for computer system design. *Journal of Systems and Software, 17*(1), 7–14.

Friedman, B., Kahn, P. H., Jr., & Borning, A. (2006). Value sensitive design and information systems: Human–computer interaction in management information systems. *Foundations, 5*, 348–372.

Grassi, A., Gaggioli, A., & Riva, G. (2009). The green valley: The use of mobile narratives for reducing stress in commuters. *Cyberpsychology & Behavior, 12*(2), 155–161.

Hekler, E. B., Klasnja, P., Froehlich, J. E., & Buman, M. P. (2013) Mind the Theoretical Gap: Interpreting, Using, and Developing Behavioral Theory in HCI Research. Proceedings of the SIGCHI Conference on Human Factors in Computing Systems 3307–3316, New York: ACM.

King, L. A., & Miner, K. N. (2000). Writing about the perceived benefits of traumatic events: Implications for physical health. *Personality and Social Psychology Bulletin, 26*(2), 220–230.

Li, I., Dey, A., & Forlizzi, J. (2010). A stage-based model of personal informatics systems. In *Proceedings of the 28th International Conference on Human Factors in Computing Systems* (pp. 557–566). New York: ACM.

Li, I., Dey, A., Forlizzi, J., Höök, K., & Medynskiy, Y. (2011). Personal informatics and HCI: Design, theory, and social implications. In *Proceedings of the 2011 annual conference: Extended abstracts on human factors in computing systems* (pp. 2417–2420). New York: ACM.

Lupton, D. (2012). M-health and health promotion: The digital cyborg and surveillance society. *Social Theory & Health, 10*(3), 229–244. doi:10.1057/sth.2012.6.

Pennebaker, J. W. (1997). *Opening up: The healing power of expressing emotions* (p. 249). New York: Guilford Press.

Pennebaker, J. W. (2004). *Writing to heal: A guided journal for recovering from trauma and emotional upheaval*. Oakland, CA: New Harbinger.

Picard, R. W. (1997). *Affective computing* (p. 275). Cambridge, MA: MIT Press.

Preziosa, A., Grassi, A., Gaggioli, A., & Riva, G. (2009). Therapeutic applications of the mobile phone. *British Journal of Guidance & Counselling, 37*(3), 313–325.

Riva, G. (2005). Virtual reality in psychotherapy (review). *Cyberpsychology & Behavior, 8*(3), 220–230.

Sellen, A., Rogers, Y., Harper, R., & Rodden, T. (2009). Reflecting human values in the digital age. *Communications of the ACM, 52*(3), 58–66.

Siegel, M. A., & Beck, J. (2014). Slow change interaction design. *Interactions of the ACM, 21*(1), 28–35.

Spahn, A. (2011). And lead us (not) into persuasion...? Persuasive technology and the ethics of communication. *Science and Engineering Ethics, 18*(4), 633–650.

Verganti, R. (2008). Design, meanings, and radical innovation: A metamodel and a research agenda. *Journal of Product Innovation Management, 25,* 436–456.

Vertegaal, R. (2003). Attentive user interfaces. *Communications of the ACM, 46*(3), 30–33.

Villalón, J., Kearney, P., Calvo, R. A., & Reimann, P. (2008). Glosser: Enhanced feedback for student writing tasks. In *Eighth IEEE International Conference on Advanced Learning Technologies* (pp. 454–458). Santander, ES: IEEE.

Weiser, M. (1991). The computer for the 21st century. *Scientific American, 265*(3), 94–104.

Yarosh, S., Radu, I., Hunter, S., & Rosenbaum, E. (2011). Examining values: An analysis of nine years of IDC research. In *10th International Conference on Interaction Design and Children. IDC 2011* (pp. 136–144). Ann Arbor: University of Michigan.

Yick, J., Mukherjee, B., & Ghosal, D. (2008). Wireless sensor network survey. *Computer Networks, 52*(12), 2292–2330.

5 A Framework and Methods for Positive Computing

Why a Framework?

Whether we grow things, fix things, make things, or provide a service, most of us believe that our work contributes to humanity in some way. In HCI, we certainly think so. We care about users; we seek to improve their experience and their lives, whether it's through increased productivity or even in hopes of affecting significant social change. For evidence, see the wide range of work that goes on in the field, from human-centered design and natural interfaces to VSD, games for health, and HCI for peace.

So if we're all working to improve lives, what differentiates positive computing? The answer lies in the field's primary goal, its measures of success, and the theoretical foundations upon which it's grounded. Whereas usability targets greater ease of use, VSD focuses on a particular set of values relevant to the project, and work in e-health aims to affect a health goal (and each of these will measure success against those targets accordingly), work in positive computing specifically targets and measures improvements to human psychological wellbeing.

Furthermore, every field will draw on theory and research most helpful to its goals (e.g., usability engineering processes, medical research for health applications, and sociocultural theory for VSD). In the same way, positive computing grounds itself in psychological wellbeing research, including the theory and methods arising from this area for measuring wellbeing and developing it.

But combining multiple disciplinary perspectives for creating new work is never easy. How do we take a model such as Richard Ryan and Edward Deci's SDT or Martin Seligman's PERMA model and introduce it into an environment of technological innovation? How do we support designers in better understanding and operationalizing wellbeing research without oversimplifying it, and how do we help wellbeing researchers better understand the potential of technology to support their work?

In the hopes that we might contribute something to this partnership between technology and the mind sciences, in this chapter we present a framework that provides a scaffold for organizing available wellbeing research and theories, linking it to tested strategies and interventions, and operationalizing wellbeing by breaking it down into actionable chunks. By "actionable chunks," we really mean those factors (such as self-awareness, gratitude, and empathy) that have been repeatedly shown in the literature to increase wellbeing. It is our hope that this framework can help form a useful bridge across disciplines that can make integration of technology design and wellbeing research occur more smoothly and successfully.

Our framework draws on the literature in multiple areas and, as such, is pragmatically designed to be flexible and inclusive of multiple psychological theories of wellbeing. Rather than arbitrarily selecting one theory, we deliberately leave room for a project's design team to identify the particular theory or theories most appropriate to their context, their goals, their disciplinary backgrounds, and the values of their stakeholders.

We then put the spotlight on methods and measures from the fields of mental health, economics, and HCI that, whether used by themselves or in combination, are among the most helpful to evaluating work in positive computing.

Targeting Conditions and Factors

To approach the larger issue of psychological wellbeing, we can make the problem more manageable by identifying both the circumstantial conditions that influence wellbeing and the determinant factors that can be cultivated to increase it.

Circumstantial conditions are those physical and environmental aspects of people's lives that have been shown to correlate with wellbeing (see Lyubomirsky, Sheldon, & Schkade, 2005). They include: personality type, multiple intelligence levels, socioeconomic variables, physical environment, relationships and family, education and life-long learning. Although the first three are important determinants, they are generally outside the sphere of positive computing's influence. The remaining three, however—namely, physical environment (which includes the digital environment), relationships, and education—are frequently mediated by technology and therefore present great opportunity.

We suggest that technology used to mediate these circumstantial conditions can be purposely designed to support determinant *factors*, within these circumstances. Determinant factors of wellbeing are psychological constructs that have been shown to have a direct relationship to wellbeing (when a factor increases, so does wellbeing and vice versa). Among them, we have elected to highlight nine in our framework—positive emotions, motivation and engagement, self-awareness, mindfulness, resilience, gratitude, empathy, compassion, and altruism—as among the most important owing to the research evidence available.

At a very basic level, we can use conditions and factors in combination to guide our design. For example, we can design the visceral, behavioral, and reflective aspects of a *digital environment* with a view to supporting a factor such as *positive emotions*. (See the chapter "Positive Emotions" in part II for specific examples.)

Of course, it's inadvisable to focus too strictly on the difference between "conditions" and "factors." To differentiate them is useful for our purposes and generally valid. However, depending on the research, some conditions may be seen as factors and vice versa.

Entry Points to Engaging in Positive Computing

In practice, a team's approach to any positive-computing project will be influenced by that team's point of entry. For example, in one hypothetical design scenario, a team might start with a factor of wellbeing—such as gratitude—and brainstorm ways that a mobile application might be used to support it. The many studies on gratitude in the positive-psychology literature would provide them with essential information with regard to optimum timing, specific strategies, and combinations of proven interventions that could inform their design. These same studies would also provide them with measures and ideas on how to best evaluate results. But starting with a factor correlated to wellbeing (such as gratitude) is just one way of engaging with positive computing.

We have seen that designers come at positive computing from multiple starting points. For example, a project might come at a design challenge from the point of view of a particular psychological theory ("How can I evaluate my interface against the components of SDT?"). Another designer may seek to create a digital version of an existing face-to-face intervention (e.g., "How can we create an app for the 'gratitude exercise'?"). Finally, positive-computing work can be triggered by evaluation results for an existing technology ("There's an alarming amount of trolling going on in our social network. Can strategies for supporting empathy be used to reduce the problem?").

Different entry points to engagement will determine how participants seek out the research and multidisciplinary partnerships that will make projects effective. All entry points benefit from access to research on determinant factors of wellbeing. We have chosen to organize the framework described in this chapter around determinant factors in part because they seem to be the most common entry point for positive-computing projects and in part because we believe that approaching the rather large and multifaceted notion of wellbeing is made more manageable by narrowing to the specific factors proven to impact it. In the framework presented here, we include those factors that have the most research literature available to support them, along with strategies and technologies already shown to develop them.

A design team might use a customization of the framework as a tool either by selecting one or more of the factors listed based on their context or perhaps by selecting a factor we have not included, but for which the literature also offers strong evidence of causal wellbeing increases.

The Framework

Our contribution to structuring the way we think about the issue of technology design for wellbeing has three major components:

1. That design for wellbeing can be managed pragmatically by focusing on one or more of its determinant factors. (We look at each of these factors in greater detail later in table 5.1.)
2. That in order to engage in design for wellbeing, we must draw factors from one or more existing theories of wellbeing and/or from significant research evidence.
3. That we can approach design for wellbeing in at least three different ways: as redesign to address or prevent detriments to wellbeing (preventative), as actively integrated promotional design (active), or as the creation of technologies dedicated explicitly to wellbeing promotion (dedicated). (We look at each of these approaches in greater detail later in table 5.2.)

Determinant Factors of Wellbeing

Column 1—Factors
Column 1 of table 5.1 contains a list of determinant factors of wellbeing. We have organized them according to their relation to the individual in

Table 5.1
Wellbeing Factors for Positive Computing

	Factor	Literature & Theory	Strategies	Methods & Measures
SELF (*Intrapersonal*)	Positive Emotions	• Hedonic psychology (Kahneman) • Subjective wellbeing (SWB) (Deiner) • Building and broadening effect (Fredrickson)	• Savoring • Positive ruminating • Reframing • Compassion meditation	Positive and Negative Affect Schedule (PANAS) Scales; general wellbeing measures such as SWB, Satisfaction with Life Scale (SWLS), and Quality of Life Scale (QoL).
	Motivation & Engagement	• Self-determination theory (Deci & Ryan) • Flow theory (Csikszentmihalyi)	• Intrinsic and extrinsic rewards	Motivation and Engagement Scale; self-regulation questionnaires
	Self-Awareness	• Cognitive behavioral therapy (CBT) (Beck) • Emotional intelligence (Mayer & Salovey)	• Life summary • Online CBT • Technology-mediated reflection	Emotional intelligence measures (e.g., Mayer-Salovey-Caruso Emotional Intelligence Test [MSCEIT]); wellbeing measures such as life satisfaction
	Mindfulness	• Mindfulness-based stress reduction (MBSR) (Kabat-Zinn) • Mindfulness-based cognitive therapy (MBCT) (Segal, Williams, & Teasdale)	• Mindfulness meditation • MBSR strategies • MBCT strategies	Mindfulness Attention Awareness Scale (MAAS); Freiburg Mindfulness Inventory
	Resilience	• Psychology of resilience (Seligman, Keyes) • Building and broadening effect (Fredrickson)	• Positive-psychology interventions • SuperBetter app	Resilience Scale

Table 5.1
(Continued)

	Factor	Literature & Theory	Strategies	Methods & Measures
SOCIAL (*Interpersonal*)	**Gratitude**	• Psychology of gratitude (Emmons & McCullough)	• Gratitude visit • Gratitude journal	Gratitude Questionnaire
	Empathy	• Emotional intelligence (Salovey & Mayer, Goleman) • Affective and cognitive empathy (Gerdes et al., Singer)	• Role playing • Perspective taking • Emotion-recognition training	Empathy Quotient; Interpersonal Reactivity Index; Children's Empathic Attitudes Questionnaire
TRANSCENDENT (*Extra-personal*)	**Compassion**	• Compassion-focused therapy (Gilbert)	• Compassion meditation	Self-Compassion Scale; Prosocial Orientation Scale; Hostile Attribution Bias Questionnaire
	Altruism	• Psychology of altruism (Batson)	• Prosocial games • Role playing helping behavior (Bailenson)	

In this table, various determinant factors of wellbeing identified in the literature are set alongside strategies for fostering them and the validated measures and scales for measuring impact to these factors.

order to facilitate linking between technology features and factors of wellbeing. Specifically, factors are placed into one of three categories: (1) self, (2) social, and (3) transcendent. We can think of the first category *self* as including those factors oriented toward "intrapersonal" phenomena. These factors are experienced within oneself, and this experience is generally not dependent on the presence of others (e.g., self-awareness and resilience). The second category, *social* factors, describes "interpersonal" phenomena, or factors that depend on interaction between *oneself* and others. (e.g., empathy and gratitude). Finally, *transcendent* factors are those extrapersonal phenomena characterized by concern and action for a greater good and for beings beyond those whom we know personally (e.g., compassion and altruism).

We have dedicated a full chapter to each of most of the factors listed in the framework, and it is these chapters (composed of background evidence, existing strategies, and design implications) that make up part II of this book. Naturally, our list is by no means exhaustive. As mentioned earlier, we made a selection based on both (*a*) the availability of research evidence for the factor's link to increased wellbeing and (*b*) the availability of technologies and design strategies to support it. Therefore, although we have included mindfulness, engagement, and self-awareness in the "self" category, that category might just as well include factors such as self-regulation, introspection, optimism, and elements of emotional intelligence.

Likewise, another of the most studied factors that would fall into what we call the transcendent or extrapersonal category is *spirituality*. According to some psychologists, a more spiritual person is more likely to be happier. The evidence is such that Christopher Peterson and Martin Seligman (2004) included spirituality in their handbook of character strengths. Despite the evidence, we are not aware of very much work studying the possible links between spirituality and technology, and for this reason we have not included it in the list or dedicated a chapter to it. Possible early exceptions include the work by Marina Bers and Seymour Papert at MIT's Media Lab (Bers, 2001) and the work reviewed by Elizabeth Buie and Mark Blythe (2013). Despite its importance to the general public (as just one example, the YouVersion of the Bible produced by LifeChurch.tv had 100 million downloads in a single month in 2013 [O'Leary, 2013]), the topic has received little academic attention, and Genevieve Bell, in her 2010 keynote address to the HCI community, mentioned spirituality, or what she called "technospiritual practices" as one of the most important underdeveloped areas of HCI (as cited in Buie & Blythe, 2013).

Other factors that might be considered transcendent phenomena include social responsibility and humility (Peterson & Seligman, 2004) as well as some of the identified components of wisdom, such as big-picture view, reflective and dialectical thinking, and an understanding of impermanence, complexity, and relativism (Calvo & Peters, 2012; Sternberg, 1990).

Column 2—Literature and Theory
In the second column of the framework (table 5.1), we include, for each factor, at least one psychological model that identifies it as important to wellbeing. Where evidence for the factor's influence is plentiful but is not explicitly implicated by a theoretical model, we reference research that demonstrates its significance.

Column 3—Strategies
In the third column in table 5.1, we list examples of strategies or interventions (technological where available) that have been tested and proven (*a*) to develop the factor and (*b*) to improve wellbeing.

Column 4—Evaluation Methodologies
In the last column of the framework, we list examples of measures and scales that have been used in the context of the factor and related interventions. These various instruments and scales are used to measure wellbeing overall or to measure the specific factor.

We feel it's critical to point out that analysis and evaluation of wellbeing will generally need to occur at a level of granularity *at least* as detailed as use case or user scenario rather than at the overall level of tool. We often lead workshops as a way to structure multidisciplinary dialogue around positive computing, where we elicit perceptions of how different user scenarios have different effects on various factors of wellbeing. For example, we have explored how using Facebook once a month to check up on remote friends might be different from using it twice an hour to communicate with friends in close physical proximity.

In other words, it's not terribly useful to speak of a social media tool as being either "good" or "bad" for wellbeing. There isn't one way to use social media and therefore not one way in which it affects wellbeing. A social network application is a system so complex that it can be both good and bad for wellbeing. Similarly, a videogame can be both good and bad for wellbeing (even simultaneously) because different aspects of it can foster prosocial behavior, while other aspects of it desensitize to violence.

A Framework and Methods for Positive Computing

The critical differences in technology's effects on wellbeing lie in different aspects of the technology, in the different ways it is used, and in different users and contexts. As such, a focus on usage scenarios (or "use cases") is a more appropriate level of granularity than is a general focus on the overall tool. The complexity with which factors and contexts are intertwined will be an important area of research for the design and delivery of new features and tools.

Other Considerations

It's also worth pointing out some considerations and ongoing challenges for evaluation. For example, factors may have different effects at different points in time. Things that increase positive emotion in the present moment may harm social relationships over time (monitor any addiction for evidence.) This consideration, combined with the potential for interventions to be fleeting, suggests that evaluations of wellbeing designs should occur not only directly after or during use, but also at later intervals in time to give evidence of genuinely beneficial long-term effect.

Moreover, it is often methodologically difficult to make a clear distinction between what is correlated with, a cause of, or a result of wellbeing. For example, people who report being happy are more likely to have positive relationships with family members and partners, are more likely to find meaning in what they do, and are more likely to be accomplished at it. But which came first? If you marry the right person, you will be happier. On the other hand, if you are a happy person, you are more likely to develop good relationships. The chicken–egg problem is one that experimental design in psychology repeatedly confronts.

Another problem is that although the factors may be correlated, they don't all have to be present or change in the same direction over time, and they may have curvilinear relationships with wellbeing. For example, we may find our work very meaningful, and it provides us with many positive emotions, yet working more does not increase our wellbeing. On the contrary, if we overwork, family relationships and personal experience suffer. This furthers the argument for evaluation over a longer term.

Integrating Design for Positive Computing into Practice

A precise definition of scope for positive computing can be honed only through maturation of the field and ongoing professional dialog. However,

Table 5.2
Approaches to the Design and Integration of Positive Computing

Approach	Integration of Wellbeing
None	Wellbeing is not systematically considered in the design of the technology (represents most technology to date).
Preventative	Obstacles or compromises to wellbeing are treated as errors and trigger redesign.
Active	Technology is designed to actively support components of wellbeing or human potential in an application that has a different overall goal (e.g., email or social networking tool designed to support users' wellbeing).
Dedicated	A technology that is purpose built and dedicated to fostering wellbeing and/or human potential in some way.

Three ways in which design for wellbeing can be and is already integrated into technology development.

we can already see a number of technology types emerging. We have used these types to identify a basic set of four integration methods, or approaches, to positive-computing design, which we list in table 5.2 and describe here to facilitate dialogue in the area. Note that this list of design types should not be seen as implying that any category of technology is superior to another. Moreover, the list is intended as a tool for communication (rather than as a rigorous taxonomy), and you will find that some technologies will fit into more than one category.

No integration. The first category might be described as the "modern baseline." It includes all of the software we have been designing for decades for which users' psychological wellbeing has not been central to specification or evaluation. As far as we are aware, most large software companies do not have a process by which user wellbeing and personal development are systematically considered, although this is beginning to change.

Preventative design. In response to the wealth of media stories on the negative impact that ubiquitous digital technologies are purportedly wreaking on our lives, our stress levels, our memories, and our children, one sensible approach to positive computing is to start by fixing evident problems. At a very minimum, we envision that in the near future, just as obstacles to ease of use and functionality are corrected during iterative stages of testing and design, impediments to wellbeing uncovered in the design process or through separate research will also be addressed as part of the design cycle. In this category, hindrances to wellbeing will be treated as errors and trigger a process of redesign.

Take, as an example, cyberbullying and trolling (the posting of inflammatory, derogatory, or misleading messages). These phenomena cause much suffering and can even contribute to drastic consequences such as suicide[1] for those affected. At the root of the problem is the *online disinhibition effect* (Suler, 2004), which has been discussed in academic literature and is an important concern for most online media outlets that accept public commentary. One design strategy for ameliorating the problem is to support users that build a reputation around their digital identities—for example, by moderating comments from new users, but allowing "trusted" users to post freely. As the user posts more comments, and if these comments are rated highly or read often, the user's reputation grows and makes it less likely that she will be willing to risk losing her status by posting defamatory comments. In this example, removing anonymity and building in reputation systems can be used as strategies to reduce the negative effects of trolling on wellbeing.

Active design. This category represents digital technologies of any kind (e.g., email, word processing, social networks) that *incorporate* functionalities and/or designs in order to *foster factors of wellbeing*, but that are not designed primarily for that purpose. Although this level of integration may take the longest to realize, subtle hints of interest in it are already bubbling up in the software industry. For example, an argument could be made that when "focus view" was added to Microsoft Word or the "distraction-free writing" button was added to WordPress, the feature was integrated for the purpose of supporting engagement and flow. In a similar way, more interface changes might be added to support empathy or gratitude in social networks or altruism in entertainment games.

Dedicated design. These technologies have wellbeing support as their primary goal. In these early stages, the most salient examples of positive computing are to be found in this category. For example, an increasing number of technologies are designed (1) to support wellbeing factors explicitly (e.g., gratitude or mindfulness apps); (2) for specific behavior change, fueled by advances in research on personal informatics and persuasive computing (e.g., goal setting and motivational apps); and (3) by mental health experts as therapeutic or promotional interventions (e.g., reachout.com).

Guidelines for Work in Positive Computing

The World Health Organization report *Prevention of Mental Disorders: Effective Interventions and Policy Options* (Hosman, Jané-Llopis, & Saxena, 2005) provides a useful summary of the state of the art, and its list of conclusions can be usefully adapted to the context of positive computing, thus providing a set of initial guidelines (see table 5.3).

As Internet-based interventions become more common, we are gathering evidence of which among them have the most positive effects. In a review of the literature on technology-supported interventions that have promoted behavioral changes to improve health (Webb, Joseph, Yardley, & Michie, 2010), the authors systematically coded common features of 85 Internet-based intervention studies with a total sample size of 43,236 participants, computing the effect size associated with each. The findings

Table 5.3
Guidelines for Positive Computing Adapted from the World Health Organization (WHO) Report on Mental Disorder Prevention

WHO Report Conclusions (Hosman et al., 2005)	Adapted to Promotion and Positive Computing
1. Prevention of mental disorders is a public-health priority.	1. Promotion of user wellbeing is a priority for technology development.
2. Mental disorders have multiple determinants; prevention needs to be a multipronged effort.	2. Wellbeing has multiple determinants; promotion needs to be a multipronged effort.
3. Effective prevention can reduce the risk of mental disorders.	3. Effective positive computing can reduce the risk of mental disorders and improve overall wellbeing.
4. Implementation should be guided by available evidence.	4. Implementation should be guided by available evidence.
5. Successful programs and policies should be made widely available.	5. Successful software and design guidelines should be made widely available.
6. Knowledge of evidence for effectiveness needs further expansion.	6. Knowledge of evidence for effectiveness needs further expansion.
7. Prevention needs to be sensitive to culture and to resources available across countries.	7. Positive computing needs to be sensitive to culture and to resources available across countries.
8. Population-based outcomes require human and financial investments.	8. Population-based outcomes require human and financial investments.
9. Effective prevention requires intersectoral linkages.	9. Effective positive computing requires involvement of all stakeholders.
10. Protecting human rights is a major strategy to prevent mental disorders.	10. Protecting human rights and other values is a major strategy to promote wellbeing.

suggest that the **effectiveness of Internet-based interventions** is associated with more **extensive use of theory**, the inclusion of **multiple techniques** for behavior change, and the use of **multiple methods of interacting** with participants (text messages were especially effective).

Methods and Measures

No matter the effort that has gone into the development of an application, its ultimate value comes down to the outcomes of its use, which makes measuring those outcomes especially critical. Identifying multidimensional measures and validated methodologies for data collection and evaluation is also essential to any funding proposal and to peer-reviewed publication. Finally, formative evaluation is essential to the ongoing process of iterative design.

But how do you measure a technology's impact on human psychological wellbeing? How can we know if a particular software or device has a positive or negative effect or any measurable effect at all on the target audience? For positive computing, the ultimate outcomes should be *measurable increased psychological wellbeing*, and there is a wealth of methodologies from HCI, psychology, and economics research to draw on, many of which have already been alluded to in preceding chapters. Although measures vary, the good news is that—as you might expect—the measures used by different researchers do agree with each other. For example, this is the case between the CES-D Scale and various measures of positive emotions. In other words, research shows that when measures for positive emotion go up, the CES depression scale goes down, and vice versa.

In the next section, we highlight a collection of these methods and measures that, by themselves and in combination, are among the most applicable to evaluating work in positive computing.

Promotional and Preventative Interventions

A report by the US National Research Council (O'Connell, Boat, & Warner, 2009) provides a taxonomy of four intervention types: treatment, maintenance (i.e., long-term management and rehabilitation), prevention, and promotion. In general, dedicated positive-computing technologies will fall in the promotional or prevention categories of interventions, and, as such, the methodologies used for evaluating mental health interventions will likely prove effective for these projects.

Promotional interventions target the wider population and are designed to develop competencies and strengths such as self-esteem, happiness, and resilience. According to the report, "Mental health promotion is characterized by a focus on wellbeing rather than [on] prevention of illness and disorder, although it may also decrease the likelihood of disorder" (O'Connell, Boat, & Warner, 2009).

Interventions for *prevention* include universal prevention (which, like promotion, targets the whole population but focuses on reducing risks), *selective* prevention (which is targeted to groups that have higher than average risk of problems), and *indicated* prevention interventions (which are targeted to high-risk individuals who have detectable symptoms indicating a likelihood of forthcoming health problems).

Interventions in mental health—be they therapeutic, preventative, or promotional—are typically evaluated using measures of "ill-being."

Measuring Ill-Being

If you consider the lack of (or lower quantity of) dysfunction or ill-being as a reasonable measure of wellbeing, then a depression or anxiety scale can serve as a useful evaluation tool. The most commonly used measure of depression is the CES-D Scale, first developed by Leonor Radloff at the National Institute of Mental Health (Radloff, 1977). The CES-D is a short self-report questionnaire designed to measure depressive symptoms, in particular affective ones such as moods, in the general population. It has been tested in household as well as clinical settings and has shown to have the required statistical internal consistency and test–retest repeatability (i.e., the measure is considered *reliable*). The *validity* of the questionnaire as a measure of depression (its ability to measure variables of clinical relevance) has been demonstrated via correlation with other self-report measures, clinical ratings, and relationship with other variables.

The CES-D consists of 20 questions, each of which is answered by self-reporting how often one has experienced the feeling indicated. The overall CES-D score is calculated giving values to each response: not at all or less than one day = 0, one to two days = 1, three to four days = 2, five to seven days = 3, nearly every day for two weeks = 4 (in some questions, the values are negative).

The Center for Epidemiologic Studies–Depression (CES-D) Scale

During the last week:
1. I was bothered by things that usually don't bother me.
2. I did not feel like eating, my appetite was poor.

A Framework and Methods for Positive Computing

3. I felt that I could not shake off the blues, even with help from family and friends.
4. I felt I was just as good as other people.
5. I had trouble keeping my mind on what I was doing.
6. I felt depressed.
7. I felt that everything I did was an effort.
8. I felt hopeful about the future.
9. I thought my life had been a failure.
10. I felt fearful.
11. My sleep was restless.
12. I was happy.
13. I talked less than usual.
14. I felt lonely.
15. People were unfriendly.
16. I enjoyed life.
17. I had crying spells.
18. I felt sad.
19. I felt that people disliked me.
20. I could not get "going."

Depending on the overall CES-D score, the diagnosis would be:

- **Meets criteria for major depressive episode.** The client has suffered anhedonia (inability to feel pleasure) or dysphoria (i.e., negative moods or sadness) nearly every day for the past two weeks. He has also reported symptoms in four other DSM symptom groups nearly every day for the past two weeks.
- **Probable major depressive episode** where the client has a score greater than 16 and meets specific criteria on symptom groups.
- **Subthreshold depression symptoms**: People who have a CES-D Scale score of at least 16 but do not meet specific criteria of low scores for specific symptom groups.
- **No clinical significance**: When the total CESD score is less than 16 across all 20 questions, the client is doing "well."

Although the CES-D seems to still be the most commonly used measure, the new version, the CESD-R (where R stands for "Revised") is increasingly applied. The CESD-R measures symptoms defined by the American Psychiatric Association's DSM-IV for a major depressive episode. Like the CES-D, it consists of 20 self-report questions that measure symptoms of depression in nine different groups:

- **Sadness** (Dysphoria): three questions
- **Loss of Interest** (Anhedonia): two questions
- **Appetite:** two question numbers
- **Sleep:** three questions
- **Thinking/Concentration:** two questions
- **Guilt** (Worthlessness): two questions

- **Tired** (Fatigue): two questions
- **Movement** (Agitation): two questions
- **Suicidal Ideation:** two questions

The CES-D has had an impressive impact not only in clinical practice, but also in academic research—more than 20,000 publications have cited it. Many of these publications have reproduced its statistical properties and its validity as a method for diagnosing depression. Owing to its history and validity, the CES-D is likely to continue to prove useful for evaluating change to overall wellbeing for many positive computing projects. However, beyond measuring the absence of ill-being, economists, as we have seen, have taken to the measure of wellbeing in both subjective and objective forms.

Measuring Wellbeing and Happiness

The seminal study conducted by Richard Easterlin (1974) discussed in chapter 3 used two types of data, a choice that has proved significant to date. The first was a Gallup Poll–type survey in which participants were asked, "In general, how happy would you say you are? *Very* happy, *fairly* happy or *not very* happy" (Easterlin 1974, p. 91, original emphasis). The second type he used was collected by psychologist H. Cantril across 14 countries. For these data, participants were asked to rate on a 1–10 scale their satisfaction with life. Happiness and life-satisfaction scales remain the most common measures used by economists today.

Life-Satisfaction and Subjective Wellbeing Scales

Happiness and life-satisfaction scales are subjective measures (i.e., self-reported) and are very different from the objective measures also used by economists. Happiness (or SWB) and life satisfaction are two separate factors; that is to say, an individual can score high in one but not on the other.

There are two further notable features of the data discussed by Easterlin: he used surveys, and these surveys represent self-reports of cognitive interpretations of happiness and life satisfaction. Of course surveys have their methodological drawbacks: they provide a "shallow" description of phenomena because participants cannot be asked follow-up questions, and the data may be prone to distortion when questions rely on interpretation or memory of events. On the other hand, the survey does allow data to be collected from a very large number of people quickly, across many countries, and at a low cost, allowing for thorough statistical analysis. As such it makes possible the huge amount of data from the Gallup World Poll, the World Values Survey, the European Social Survey, and the *World Happiness Report* (Helliwell, Layard, & Sachs, 2012).

The measures of SWB used by economists have been tested for reliability in the same way that mental health measures have. At the individual level, for example, life evaluations performed as a sequence of surveys start with a high correlation, and as the surveys get farther apart in time, this correlation gets lower. At the national level, we see the same phenomenon: year-to-year correlations start at high values (0.88 to 0.95) and are reduced across longer time spans reflecting long-term change (Helliwell et al., 2012).

Objective Utility versus Remembered Utility
The questions asked in wellbeing surveys can be designed to elicit "instantaneous," moment-by-moment reports or "retrospective" reports, for which a participant is asked to remember how he or she felt at another point in time. Each has advantages and disadvantages. In the former, you have to reach the participant and interrupt what she is doing to ask the question (something both technically difficult and possibly aggravating for the participant). For the latter, since we are not great judges of how we felt in the past, the researcher has to take into account the factors influencing appraisal and recollection of events.

The time span covered by a report is also significant: "How do you feel now?" or "How did you feel at noon?" are different from "How did you feel this week?" Both ways of asking the question would lead to different types of answers. In a landmark study surveying participants of a medical procedure, cumulative reports of moment-by-moment pain were not equivalent to retrospective summary reports of the pain experience. (Redelmeier & Kahneman, 1996). The study showed that the experience of pain at the end of the procedure disproportionately affected memory of the procedure as a whole. We discuss this peak–end rule further in the chapter on positive emotions.

Daniel Kahneman (1999) has formalized a measure of "objective happiness" that averages the moment-to-moment emotional state of a person over time. This measure is based on the concept of *instant utility*, which states that every event or action will either be good in our view (we want to keep doing it) or bad (we want to interrupt it) in the moment it occurs. If we ask a person to make a retrospective evaluation of that same event, that valuation may be different and is referred to as *remembered utility*. Objective happiness is frequently measured via experience sampling.

Experience Sampling versus Automated Detection

Kahneman's approach of asking for instantaneous experience reports and then averaging them leverages the Experience Sampling Method (Csikszentmihalyi & Larson, 1992). This method has become increasingly common given the availability of mobile phones (via notifications or text messages), which can be used to interrupt a person's activity to ask a few questions. This technique was used in a study reported in *Science* by Matthew Killingsworth and Daniel Gilbert (2010) on the impact of mind wandering on happiness, which we discuss in more detail in the chapter on mindfulness.

The problem with experience sampling is that it is intrusive because participants have to be interrupted frequently. Back in 1999, Kahneman (1999) already contemplated possibilities for solving this problem by suggesting that information about affective states might be derived from noninvasive sources such as brain signals. Nowadays this is a goal of affective computing, which has produced a body of research on different automated affect-detection techniques. Affective-computing techniques could make the process of collecting data about wellbeing much easier and ideally more accurate in the near future.

One modern-day example moving in this direction is called Mood-Meter, a computer vision tool that counts smiles (Hernandez, Hoque, Drevo, & Picard, 2012). The system was created as a demonstration for a special event at the MIT Media Lab and was evaluated over a period of 10 weeks, during which time people walking by the installation were able to interact with it. The system recognized and counted user smiles. The authors uncovered periodic patterns, such as more smiles during weekends and fewer during exam periods. These results correspond with the idea that smiles are correlated to positive emotions, and they could prove a helpful way of estimating user experience of positive emotions in the future. The recognition of facial expressions that correspond to negative emotions could provide another nonintrusive way to estimate hedonic values.

Human–Computer Interaction Methods

Historic accounts of how HCI and its measures have evolved (e.g., Preece et al., 1994) often begin by describing how research in the 1980s focused on office automation software that had very simple interfaces, and researchers analyzed human and operational error. In the early 1990s, interest shifted toward new Internet-based communication technologies such as email, chat, and groupware, and toward improving and measuring ease of use. In the early 2000s interest extended into web 2.0 applications and the notion of a user experience hit the mainstream along with methods for evaluating it. Now we have measures for a variety of aspects of user experience (such as usability, value, and pleasure) but not specifically for wellbeing. As such, here we look at general HCI methods that can be used in positive computing work, as well as at measures that have been used both in positive psychology and to evaluate Internet-based mental health programs that can be incorporated into HCI in order to evaluate and inform the development of positive-computing interventions.

Self-Reports, Observation, and Experimentation

You can group current available research methods into three basic types. First, there are ***descriptive investigations*** such as the Gallup-type surveys mentioned earlier and ***observation*** methods such as focus groups and interviews, both of which allow researchers to better understand a population as a whole. Data collected in these two ways can be used to understand how one factor relates to another in what is called relational investigation. But these two approaches are seldom sufficient to isolate specific causes with certainty. To study causality, researchers use ***experimental designs*** (e.g., randomized controlled trials or multivariate testing). These three methodologies are often used in combination within the same project and may have significant overlap.

Sometimes experimental designs are not possible due to cost, time, or even ethical reasons. Imagine you are a manager in a company and would like to test the impact of a longer maternity-leave allowance on employee wellbeing. The cost of such a policy change can be significant, and the outcomes may only be measurable over years. Furthermore, it is not easy to understand causality. If we do measure a change to employee wellbeing, it would be hard to separate out other causes associated with parenthood. In order to understand causality, we need controlled experiments, but there are ethical implications to such experiments: Could we give benefits to just half of staff and use the other half as a control group? How would you select those who would receive the better deal?

When experimental designs are possible, they can inform the development of software applications and digital artifacts. Experiments in HCI are, at their core, the same as any other experiment in physics or medicine.

They start with a hypothesis: a precise statement that can be tested directly through a single experiment and that clearly specifies the control groups and conditions in which the experiment is to be executed. One common type of experiment involves comparing two designs. In this case, the hypothesis is generally divided in two: the null hypothesis H_0, which assumes there is no *significant* difference between the two, and the hypothesis H_1, which assumes there is.

Take, for instance, two hypotheses for a controlled positive-computing experiment comparing two designs or interventions:

- H_0: There is no difference between design A and design B on SWB measures.
- H_1: There is a difference between design A and design B on SWB measures.

These two hypotheses would have emerged from previous research, and part II of the book discusses research specific to possible design strategies.

A hypothesis always identifies an independent and dependent variable, along with the factors that cause the change in the dependent variable. For positive-computing applications, the dependent variable would generally be one of the measures of wellbeing, and the independent variables would be new designs or features. In some cases, the dependent variable will be a factor for which there is strong evidence of support for wellbeing (e.g., gratitude).

In psychology, experimental designs follow the same methodological rigor as medical interventions. Many studies in both psychiatry and positive psychology use the CES-D Scale to measure depression/wellbeing. Among the most prestigious journals of experimental research at the intersection of psychology and technology is the *Journal of Medical Internet Research*, which publishes research (including randomized controlled trials) of different Internet-based interventions. In these studies, one or more treatments are evaluated using the treatment as the independent variable and the CES-D measure (or something similar) as a dependent variable.

One such study (Schueller & Parks, 2012), compared combinations of six types of self-help positive-psychology exercises:

- **Active-constructive responding**: Practices to respond positively to good news shared by others, something that many have trouble with. This can be done by asking a participant to relive an experience.

- **Gratitude visit**: An exercise first studied by Seligman in which a participant writes a letter of gratitude and then reads it aloud to the person being thanked.
- **Life summary**: An exercise also studied by Seligman in which a participant is asked to write a summary of how he would like his kids to remember him.
- **Three good things**: In a diary-type exercise, the participant records three things that went well each day along with reasons.
- **Savoring**: The participant is asked to focus on a positive experience two to three times each day.
- **Strengths**: The participant receives individualized feedback about her strengths based on responses to Seligman's Values-in-Action Strengths Questionnaire.

The Schueller and Parks study is particularly useful for understanding how design decisions for wellbeing software might be made. Within the context of the psychology literature, the study's main contribution is an evaluation of multi-exercise packages (as opposed to the more commonly studied single-exercise approach). The researchers recruited 1,364 self-help-seeking people who were randomly allocated to one of three designs: two, four, or six online activities (or assessments in only the control group) over a six-week period. These exercises drew from the content of group positive psychotherapy. Participants interacted with a website that provided instructions, sent email reminders, and contained the baseline and follow-up assessments. The evaluation has similarities with those common in HCI research where after participants completed the exercises for each design being evaluated, they reported their enjoyment of it, answered how often they had used it, and completed a questionnaire that measured the impact (CES-D).

Perhaps the most important result was the efficacy of the treatments: all the designs produced significant reductions in depressive symptoms ($F_{1,656}$ = 94.71, $P < 0.001$). The efficacy was measured using a repeated measures analysis of covariance on the measures of depressive symptoms. When participants started the treatment, their average scores on the CES-D Scale indicated mild to moderate levels of depressive symptoms (with a mean of 16.93 and standard deviation of 12.64, 16 being the minimum for a clinical depression). The individuals in each design group (packages with two, four, and six exercises) and the controls started with similar levels of depressive symptoms.

Also important (especially for designers) is that the study showed a significant difference for the impact of *different designs*. The improvement, measured as a reduction in a depression measure, was larger for the groups that received two or four exercises compared to those receiving the six-exercise package or the control condition. In statistical terms, the study showed a significant condition by time interaction ($F_{3,656} = 4.77$, $P = 0.003$).

While this particular study sheds light on a specific design issue (the optimal number of exercises for one program), wellbeing research is often exploratory, pointing to factors known to influence wellbeing yet not going so far as to explain which designs may be better at developing any particular attribute. Some of these studies tell us that people who measure highly for A (e.g., resilience) self-report highly for B (e.g., life satisfaction). Sometimes they tell us that those who, for example, participate in an intervention C (e.g., a mindfulness course) score higher on measure A (e.g., life satisfaction) and lower on B (e.g., depression scale). In positive computing, we can use this research to guide hypotheses and designs and then test those designs experimentally.

Ethnographic Study and Automated Data Collection

Ethnographic studies common in HCI research take their lead from anthropology. For ethnographic studies, a researcher embeds herself in a community in order to understand specific phenomena through observation and direct interaction. Ethnographic research allows researchers to collect a great deal of detail about the interactions and activity being investigated and allows data to be collected from within authentic settings. It is generally inductive, so the data collected in the study (often audio and video recordings) are used to identify patterns and often to produce theories that explain those patterns. Formal analysis is then used to produce ethnographic reports, which often contain coding schemes, taxonomies, models, and/or narratives for which the specific situations have been abstracted. Ethnographic research has provided in-depth information about numerous HCI problems and can be considered part of the HCI researcher's standard tool kit.

Ethnographic research on ubiquitous technologies highlights some of the increasing challenges of user-experience research. Newer and increasingly embedded technologies are adaptive, personalized, and highly distributed. Interactions occur synchronously and asynchronously, online, and/or on mobile devices with small displays that make video recording difficult (Crabtree et al., 2006). Furthermore, users are constantly interacting with multiple invisible systems, such as GPS, that influence behavior and make interaction harder to understand out of context. Finally, interactions can occur across multiple applications (e.g., multiple browsers) and devices (e.g., laptop and tablet).

The challenge posed by ethnographic research is that it is very time consuming to conduct and often even impossible to carry out. For example, it's very difficult for researchers to observe the use of technology in family life authentically over extended periods of time.

A less time-consuming approach, albeit limited in other ways, is provided by automated methods. Because we are studying how people use technologies, these technologies themselves can be used for data collection. This has been done in multiple studies—for example, using MoodMeter to read the facial expressions of passers-by (Hernandez et al., 2012), using Tweets as simple proxies for what is happening in people's lives and for inferences about their states of mind (Golder & Macy, 2011) or using the mobile phone app EmotionSense (Rachuri et al., 2010) to collect behavioral data along with emotional self-reports. The powerful potential of an automated approach is seen in recent studies by Microsoft Research that demonstrate the ways in which public Twitter and Facebook data can be used to predict and characterize postpartum depression (De Choudhury, Counts, Horvitz, & Hoff, 2014).

Other automated methods used for tracking user behavior, affect, and cognitive states across multiple applications include the use of web logs, where the proxy or browser stores information about visited websites. A more sophisticated approach to collecting data automatically is through the customized modification of software, referred to as *instrumented software*, which allows interaction to be recorded by bespoke measuring tools. The software might, for example, be modified to record every action (i.e., menu choices and keystrokes) with a time stamp. When the researcher analyses these data, he can make inferences about which features are used frequently, which ones rarely, and which for extended or short periods of time.

The area of behavioral analytics engages in studying the large amounts of behavioral data we collect from large numbers of people, sometimes in order to provide feedback to the user, for research, for marketing, or for other purposes.

In addition to the technical challenges of recording and then analyzing big data, automated data collection has its methodological disadvantages. Making sense of large quantities of data can be challenging and the results generally do not provide information on *why* people do what they do, which is why combining methods may be the best approach for research in positive computing.

Combining Methods for the Best of all Worlds

The HCI community's experience studying ubicomp environments shows that it's best to combine data collected by traditional ethnographic methods (e.g., video) with data coming directly from the digital environment, such as screen recordings or log files. Take, for example, *Savannah* (Crabtree et al., 2006), a role-playing educational game in which players learn about the behavior of lions. Players, in groups of six, use handheld devices with WiFi and GPS capabilities. They pretend to be lions, moving around on the playground while exploring a virtual savannah that appears overlaid on their devices. They explore the landscape of the savannah and must discover the resources that lions need to survive. In this study, researchers combined video and microphone recordings of the players' movements and conversations, in the ethnographic tradition, with GPS movement data and log files from the game. This approach is similar to the combination of qualitative and quantitative methods used in the social sciences and recommended for the study of wellbeing (Camfield, Crivello, & Woodhead, 2009).

Besides cross-validation, combining multiple methods allows for the various multidimensional aspects of the wellbeing construct to emerge independently (Felicia Huppert, director of the Cambridge Well-being Institute and government adviser on wellbeing measures, discusses this multidimensionality in her sidebar in chapter 2.) The data collected by combining more than one research methodology can produce a much more complete description of what is actually happening during the interaction.

We look at more examples of combination methods and multidimensional methods in the next half of the book within the contexts of various wellbeing factors. Now let us head to part II, in which we zoom in on specific factors of wellbeing included in the framework, summarizing some of the most significant research surrounding each, how each factor relates to greater wellbeing, how it has already been incorporated into technology designs, how it can be evaluated, and implications for future work. See you in part II.

Note

1. Some cases that received media coverage include the suicides of Katie Webb (Canada), Megan Meier (United States), and Hannah Smith (United Kingdom). Sadly, coverage of many more cases can be found online.

References

Bers, M. U. (2001). Identity construction environments: The design of computational tools for exploring a sense of self and moral values. PhD diss., MIT.

Buie, E., & Blythe, M. (2013). Spirituality: There's an app for that! (But not a lot of research). In F. Paris (Ed.), *ALT.CHI 2013* (pp. 2315–2324). New York: ACM.

Calvo, R. A., & Peters, D. (2012). Positive computing—technology for a wiser world. *Interactions, 19*(4), 28–31.

Camfield, L., Crivello, G., & Woodhead, M. (2009). Wellbeing research in developing countries: Reviewing the role of qualitative methods. *Social Indicators Research, 90*(1), 5–31.

Crabtree, A., Benford, S., Greenhalgh, C., Tennent, P., Chalmers, M., & Brown, B. (2006). Supporting ethnographic studies of ubiquitous computing in the wild. In *Proceedings of the 6th Conference on Designing Interactive Systems* (pp. 60–69). New York: ACM.

Csikszentmihalyi, M., & Larson, R. (1992). Validity and reliability of the experience sampling method. In M. deVries (Ed.), *The experience of psychopathology: Investigating mental disorders in their natural settings* (pp. 43–57).Cambridge, UK: Cambridge University Press.

De Choudhury, M., Counts, S., Horvitz, E., and Hoff, A. 2014. "Characterizing and Predicting Postpartum Depression from Shared Facebook Data." In *Proceedings of the 17th ACM conference on Computer supported cooperative work & social computing—CSCW '14* (pp. 626–38). New York: ACM.

Easterlin, R. A. (1974). Does rapid economic growth improve the human lot? Some empirical evidence. In P. A. David & M. W. Reder (Eds.), *Nations and households in economic growth: Essays in honor of Moses Abramovitz* (vol. 8, pp. 88–125). New York: Academic Press.

Golder, S. A., & Macy, M. W. (2011). Diurnal and seasonal mood vary with work, sleep, and day length across diverse cultures. *Science, 333*(6051), 1878–1881.

Helliwell, J., Layard, R., & Sachs, J. (2012). *World happiness report*. New York: Earth Institute.

Hernandez, J., Hoque, M. E., Drevo, W., & Picard, R. W. (2012). Mood meter: Counting smiles in the wild. In *Proceedings of the 2012 ACM Conference on Ubiquitous Computing* (pp. 301–310). New York: ACM.

Hosman, C., Jane-Llopis, E., & Saxena, S. (Eds.). (2005). *Prevention of mental disorders: Effective interventions and policy options*. Oxford: Oxford University Press.

Kahneman, D. (1999). Objective happiness. In D. Kahneman, E. Diener, & N. Schwarz (Eds.), *Well-being: The foundations of hedonic psychology* (pp. 3–25). New York: Russell Sage Foundation.

Killingsworth, M. A., & Gilbert, D. T. (2010). A wandering mind is an unhappy mind. *Science, 330*(6006), 932.

Lyubomirsky, S., Sheldon, K. M., & Schkade, D. (2005). Pursuing happiness: The architecture of sustainable change. *Review of General Psychology, 9*(2), 111–131.

O'Connell, M. E., Boat, T., & Warner, K. (2009). *Preventing mental, emotional, and behavioral disorders among young people: Progress and possibilities*. Washington, DC: National Academies Press.

O'Leary, A. (2013). In the beginning was the word; now the word is on an app. *New York Times*, July 27.

Peterson, C., & Seligman, M. E. P. (2004). *Character strengths and virtues: A handbook and classification* (p. 800). Oxford: Oxford University Press.

Preece, J., Rogers, Y., Sharp, H., Benyon, D., Holland, S., & Carey, T. (1994). *Human–computer interaction*. Boston: Addison-Wesley.

Rachuri, K. K., Musolesi, M., Mascolo, C., Rentfrow, P. J., Longworth, C., & Aucinas, A. (2010). EmotionSense: A mobile phones based adaptive platform for experimental social psychology research. In *Proceedings of the 12th ACM International Conference on Ubiquitous Computing* (pp. 281–290). New York: ACM.

Radloff, L. S. (1977). The CES-D scale: A self-report depression scale for research in the general population. *Applied Psychological Measurement, 1*(3), 385–401.

Redelmeier, D. A., & Kahneman, D. (1996). Patients' memories of painful medical treatments: Real-time and retrospective evaluations of two minimally invasive procedures. *Pain, 66*(1), 3–8.

Schueller, S. M., & Parks, A. C. (2012). Disseminating self-help: Positive psychology exercises in an online trial. *Journal of Medical Internet Research, 14*(3), e63.

Sternberg, R. (1990). Wisdom and its relations to intelligence and creativity. In R. Sternberg (Ed.), *Wisdom: Its nature, origins, and development* (pp. 142–159). Cambridge, UK: Cambridge University Press.

Suler, J. (2004). The online disinhibition effect. *Cyberpsychology & Behavior, 7*(3), 321–326.

Webb, T. L., Joseph, J., Yardley, L., & Michie, S. (2010). Using the Internet to promote health behavior change: A systematic review and meta-analysis of the impact of theoretical basis, use of behavior change techniques, and mode of delivery on efficacy. *Journal of Medical Internet Research, 12*(1), e4.

II

6 Positive Emotions

In preschools across the globe, scampering boys and girls paste their scribbly creations into "Wow" books. "Do you know why it's called a 'wow book'?" my daughter asks me (Dorian) one afternoon, eager to divulge the surprise. "Because it makes you go 'Wow!'" Her gleeful answer may not have been a total revelation, but it does reveal something more profound than might be expected. This simple handcrafted artifact is designed to support shared moments of positive emotion: joy, pride, connectedness, awe, and love. By offering an opportunity for the experience of positive emotion (both for the children and for those with whom they share it), the wow book not only increases experienced happiness but also contributes, in its own way, to flourishing (and we will find out how in this chapter). Surely it's not too much to ask for our technologies to do the same.

Positive emotions are among the few factors of wellbeing that designers already consciously attend to. Marketers try to manufacture the "wow" factor for its profit potential, and the designer among us (Dorian) can attest to the fact that we love to evoke feelings such as delight, pleasure, fun, and satisfaction through our work. This shouldn't be taken for granted in the technology industry, however. Only a decade ago technology design was lorded over by the negative, with lists of usability "violations" and an unrelenting focus on efficient function. In response, many professionals looking for the more holistic perspective offered by design in other fields protested. Naked function wasn't enough, and ease of use would not be the end of the story.

Thanks to industry pioneers such as Don Norman, Peter Wright, and Marc Hassenzahl, we have taken part in an era of human-centeredness that shifts focus from the technology to the user's experience of it. Today, when we design digital environments, we seek to create conditions for positive user experiences, knowing that positive emotions will not only be more rewarding for the user, but also be far more effective at meeting business goals, whether that's by communicating better, increasing sales, supporting learning, increasing loyalty, or meeting any of the other common objectives served by our software and tools.

So what if we could direct this practice and momentum higher, toward positive computing? What if we designed for a wider range of positive emotions not primarily because they support business goals, but as a means of increasing the psychological wellbeing of people? What research could we draw on, and where could we look for examples? To explore these questions, in this chapter we turn to some of the seminal research on positive emotions by leading psychologists such as Barbara Fredrickson, Sonja Lyubomirsky, Don Norman, and Daniel Kahneman. We also spotlight some of the technologies already designed to promote wellbeing by increasing positive emotions, inferring possible design implications and how we might extend current work into the digital future.

When Positive Emotions Increase Wellbeing

Hedonism and Wellbeing

We know positive emotions such as pleasure, serenity, and joy feel good as we experience them, but we also lament their transience. Therefore, how can they be anything but fleeting and casual contributors to our lasting happiness? Can we reconcile these ephemera to a more stable notion of long-term wellbeing?

The classic hedonist Aristippus argued that it is pleasure we actually seek in everything we do. A logical follow-through might then be to spend our days seeking to increase the number and duration of pleasurable emotions we feel, aiming for as close to constant pleasure as possible. As such, technologists could fulfill a wellbeing mission simply by creating ever more exciting games and facilitating the distribution of funny cat videos.

Without detracting from the value of cat videos, implying that pleasure suffices for wellbeing is a notion that has rarely satisfied in the history of philosophy and psychology. A constant pleasure-seeking approach to life (in addition to sounding fairly exhausting) manifests as decidedly self-defeating in cases such as addiction. However, even assuming we qualify this quest, as Epicurus did, by requiring that we take into account the effects of current pleasure-based acts on future happiness, the approach still feels inadequate. Perhaps it fails to inspire because it leaves our potential for wellbeing largely to circumstance—to the slings and arrows of outrageous fortune that will inevitably bring either pleasure or pain. Those of us not totally convinced by a desire-satisfaction model of happiness are certainly not alone, and it's from other philosophers in ancient Greece that we have inherited more complex notions of happiness, such as eudaimonia.

Nevertheless, hedonism has maintained a thriving career, and perhaps that is because we have little choice in the matter. The utilitarian Jeremy Bentham suggested hedonic enslavement as a point of departure for economic theory: "Nature has placed mankind under the governance of two sovereign masters, pain and pleasure. It is for them alone to point out what we ought to do, as well as to determine what we shall do."[1] Are we biologically predestined to seek pleasure and avoid pain as a consequence of being animal? At the turn of the twentieth century, behaviorists such as Ivan Pavlov and B. F. Skinner produced ample evidence for this conclusion via myriad lab experiments and proposed it to explain all human behavior.

Modern research on positive emotion, however, has provided us with a far more nuanced and sophisticated perspective that goes well beyond stimulus-response and gives evidence for both agency with regard to our experience of positive emotions and their potentially profound impacts. Work such as Barbara Fredrickson's in the area of positive psychology has advanced theories that explain not only why positive emotions have been essential from an evolutionary standpoint, but also how they might trigger lasting flourishing.

The Broadening and Building Effect

Negative emotions such as fear, anger, and disgust have long been explained by their importance in motivating useful survival actions, including escaping danger, fighting threat, and expelling poison. In essence, these emotions cause us to narrow in on a specific set of behavioral options most suitable to surviving a life-or-death situation.

In her seminal work, Fredrickson (2001) tackles the less straightforward survival benefits of specific positive emotions. She argues that while negative emotions narrow our options in order to facilitate quick survivalist decision making, positive emotions do the opposite. "The positive emotions of joy, interest, contentment, pride, and love appear to have a complementary effect: They broaden people's momentary thought–action repertoires, widening the array of the thoughts and actions that come to mind."

In other words, positive emotions broaden our behavior options at those times when fight or flight is not the imperative and allow us to take advantage of these moments of relative safety to think more creatively, do things differently, innovate, and build our resources (she identifies physical, intellectual, social, and psychological resources). This investment in resource building increases our chances of survival over the *long* term. To a digital gamer, this is like switching from survival mode to creative mode in *Minecraft* (and, indeed, games arguably present the richest and broadest set of examples of technology support for positive emotion to date).

Fredrickson reviews experimental findings of phenomenologically distinct positive emotions that are linked empirically to evolutionarily advantageous behaviors. Specifically, research has demonstrated a causal link between

- **joy** and an urge for playfulness and creativity;
- **interest** and an urge to explore and learn;
- **pride** with an urge to share news of achievements with others and envision greater progress;
- **contentment** and an urge to reflect and integrate savored experience into new worldviews and self-views; and
- **love**, which Fredrickson (2001) describes as "an amalgam of distinct positive emotions (e.g. joy, interest, contentment) experienced within contexts of safe, close relationships [that] broadens by creating recurring cycles of urges to play with, explore, and savor experiences with loved ones."

There are many others whose work supports these conclusions. Alice Isen (1990) suggests that positive affect leads to "broad, flexible cognitive organization and ability to integrate diverse material." Moreover, studies abound on how happy moods lead to more open-ended thinking and effective problem solving. Sonja Lyubomirsky, Laura King, and Ed Diener (2005), for example, show how positive moods facilitate a variety of approach behaviors and positive outcomes.

But according to research, there's even more to positive emotions than adaptively significant resource building. Veteran positive-psychology researchers Ed Diener and Robert Biswas-Diener (2008) identify positive emotions as critical to what they call "psychological wealth." They explain that "Happiness is the fundamental building block of psychological wealth, but it is important not just because it feels good, but because it is so beneficial in so many areas of life. Rethinking happiness requires us to understand that it is not just a pleasant goal, but necessary to achieving success in many domains of life."

Likewise, Martin Seligman includes "positive emotions" as one of the five pillars of flourishing. Fredrickson's work goes furthest to support this claim. She has amassed compelling evidence to support the idea that the experience of positive emotions in the right quantity can trigger an "upward spiral" of psychological flourishing.

The Tipping Point
Based on her work with Marcial Losada and others, Fredrickson concludes that positive emotions, in a quantity beyond a certain "tipping point," is consistently predictive of psychological flourishing. In other words, if your positive emotions surpass your negative emotions by about three to one, research predicts you will exhibit the characteristics of someone who is experiencing optimal mental health and wellbeing (Fredrickson & Losada, 2005). The idea is a parallel to the more conspicuous negative tipping point beyond which negative emotions can spiral into clinical depression.

Because Fredrickson's findings show that just as negative emotions can spiral into depression, positive emotions can spiral into flourishing, positive emotions are evidently worth cultivating not only for their own sake, but also because they generate stable mental health. SuperBetter (invented by Jane McGonigal) is currently the best example of a technology (we would call it positive computing) that is designed around the notion of and research behind psychological resources. SuperBetter is a mobile app and web-based environment that pragmatically supports small acts of wellbeing in a variety of goal contexts.

Positive Emotions versus Positive Thinking: The Vital Importance of Authenticity
Although supporting the positive is clearly valuable for wellbeing, should the benefits of positive emotions lead us to turn all bad thoughts to good, avoid difficulty, and dodge negativity in any way we can? Surely this would get us to the holy grail of a three-to-one ratio most efficiently? Yet dedicating a slew of tools intended exclusively to "cheer people up" feels suspiciously limited as an overarching goal for positive computing, perhaps because we instinctively sense the tendency for any such endeavor to lack authenticity. In fact, research has demonstrated how optimal functioning relies on the experience of negative emotions *as well*, which both have evolutionary purpose and distinguish authentically happy human beings from the fictional Pollyanna and the nonfictional sociopath.[2] In their paper on the positivity ratio, Fredrickson and Losada (2005) conclude that "appropriate negativity is a critical ingredient within human flourishing that serves to maintain a grounded, negentropic system" and that "the

complex dynamics of flourishing first show signs of disintegration at a positivity ratio of 11.6." In other words, positive feelings *can* go too far.

In fact, authenticity of emotion is so critical to wellbeing that "faking" happiness by repressing negative emotion and affirming positive thoughts in the face of negative experience have been shown to be not just ineffective, but actually harmful to one's wellbeing.

Psychology writer and *Guardian* journalist Oliver Burkeman (2013) dedicated an entire book to the importance of the negative. Although "reframing," that stalwart of cognitive behavioral therapy, is proven to be highly effective, there is an important line to be drawn between reframing and denial or repression. Considering alternate interpretations is different to either ignoring the possibility of negative events or denying the existence of your negative emotions. Burkeman points to Western business culture's tendency to repress any consideration of negative outcome or failure as a contributor to the 2008 financial crisis.

The harmful effects of inauthenticity can become a chronic problem for people who have to fake happiness as part of their job. "Emotional labor," a term introduced by Arlie Hochschild (2003) in the 1980s, refers to the effort made by employees who must display certain emotions as part of their work (e.g., sales people, waiting staff, nurses, actors). Emotional work is becoming increasingly common in the ever-growing services industry, and if it is not properly managed, it can lead to burnout and depression. Researchers have developed management strategies such as support groups and psychological work breaks where employees can "be natural."

It is in our interaction with others that we observe another case for the importance of negative emotions. Meaningful connection with others is critical to wellbeing, and it involves expressing our emotions and empathizing with others. These acts by definition require us to share in others' experience of negative emotions. We delve deeper into the wellbeing links to empathy in chapter 10.

The sophisticated dance between negative and positive emotion is easy to see when we play games. Positive experiences (such as flow) are contingent on "appropriate challenge," which means coming up against stressful and frustrating but surmountable limits and obstacles. If the obstacles weren't challenging or the limits not somewhat frustrating, there would be no rewarding sense of joy and pride in overcoming them.

Why is this important for technologists? One can easily imagine a well-intentioned design team looking to motivational speakers or self-help writers who push an imbalanced positive-thinking approach in order to inform their design work. Seen as a moral tale, the critical caveats surrounding the role of positive emotion for wellbeing highlight the importance of engaging with academic psychologists rather than being satisfied with pop psychology or trends. Where there is plenty of talk but little research evidence, not only can resources be wasted, but actual harm can be done. Again, the imperative for interdisciplinarity is clear.

In practice, few engineers will be aiming to build happiness-generating machines. Instead, design will likely find most benefit from viewing positive emotion in the context of the project and its requirements—in other words, by applying those specific strategies best suited to the activities and contexts being supported. For example, supporting gratitude in social networks, fostering serenity in a mindfulness app, or supporting interest in a productivity tool will generally be more practicable than aiming vaguely at positive emotion in general.

Reality versus Memory

The epic battle between reality and memory is an ancient and powerful one that takes place before us every day. These two competing selves (as Kahneman [2011] puts it), our "experiencing self" and our "remembering self," frequently go head to head, and the results are anything but rational. Experiments frequently rely on "self-reports" after an experience, but how good are we really at remembering what emotions we have experienced in the recent past? Apparently, not that good, which may or may not be a problem, but it certainly raises some interesting questions for designers of digital experiences.

Take, for example, the famed colonoscopy study (Redelmeier & Kahneman, 1996), which turned our assumption that human beings are rational decision-making agents on its head. In a nutshell, patients who had longer procedures (that entailed more overall pain but ended on a less painful note) reported better experiences than patients who had shorter procedures that therefore involved less overall pain, but that ended at a more uncomfortable point.

Although the digital experiences we design are, fortunately, seldom as uncomfortable as colonoscopies were before anesthesia, the very same phenomenon has been demonstrated in other ways. One simple study showed that people will willingly choose to listen to an annoying sound for longer *if* the extra length means the listening experience ends more peaceably (described in Kahneman, 2000).

Such studies provide strong evidence for what is now referred to as a "peak–end rule," which shows that our memories of events are not aggregations of our moment-by-moment experience, but rather a *reconstruction* based on what we felt at the *peak* of the experience and what we felt at the *end* of it. This has its parallels in our working-memory capacity, which remembers beginnings and ends of information lists more easily. Don Norman has suggested that in light of these facts about human psychology, it's more important to design for remembered experience than to focus on perfecting moment-by-moment interaction (see his sidebar in this chapter for more detail).

Positive Emotions Are Not Created Equal

Virtually every one of the 5,000 advertisements we see in a day (Story, 2007) is designed to get us to *want* something. Our brain gets plenty of training in desire and striving, and we're conditioned to think that acquiring things will make us happy. We are less reminded to value those things that actually are more effective at generating long-term happiness, such as social connectedness, mindfulness, and engagement. Yet both acquiring things and connecting to people come packaged with positive emotions, so what's the difference, and does it matter which positive emotions we support?

Paul Gilbert and Choden (2013) describe two physiological systems linked to two groups of positive emotions, each of which is distinguished by distinct evolutionary, neurological, and physiological features:

1. **The excitement and drive system.** Associated with the neurotransmitter dopamine and with activation of the sympathetic nervous system, the excitement and drive system generates states such as pride, enthusiasm, and desire as well as urges linked to achievement, acquisition, consumption, and ownership. From an evolutionary standpoint, this system energizes and rewards us for seeking out food, sex, shelter, and other good stuff.

2. **The affiliative and soothing system.** Associated with the release of endorphins and possibly oxytocin as well as with activation of the parasympathetic nervous system, the affiliative and soothing system generates states such as calm, compassion, and love and urges us to look after, prevent harm to, and care for others, as well as see them flourish. From an evolutionary standpoint, it motivates and rewards us to look after offspring, form and tend to social relationships, and take care of the world we live in.

Whereas the drive system gives us the hyped-up buzz that follows achievement or accompanies anticipation of reaching a goal (so familiar in gameplay), the affiliative system gives us the warm glow of affection, the fulfilling joy of savored experience, and a great time out with friends. Both are important, and, naturally, they frequently come in combination. But it is the affiliative system that is critically necessary for tempering our negative emotions (such as anger and fear) and for keeping the excitement and drive system in check. Unchecked, the drive system can lead to selfish determination and destructive pleasure seeking, greed, addiction, and loneliness. As Gilbert and Choden (2013) put it, "Harnessed up, drives are helpful and essential, but unregulated they can be extremely damaging. Excessive drive leads to over self-focus, immoral and corrupt practices because of the enticements," as well as "callousness and indifference towards those who suffer."

Gilbert and Choden also explain that the affiliative system keeps us in balance physically and mentally. "Our brains are set up to be calmed down in the face of kindness. … [K]indness and feeling connected to people will help balance your sympathetic and parasympathetic nervous systems—and this can be the case whether the kindness and affiliation comes [sic] from yourself or from those around you." Gilbert and others have actually found that compassion and self-compassion are sources of resilience and mental flourishing, and their cultivation is effective in the treatment of mental health problems (but we get to that in chapter 12).

In the context of design for positive emotions, the difference between these two systems and the consequences of an imbalance should lead us to look at the various ways our technologies impact each of them. To our knowledge, there has not been any research in HCI from this perspective.

We would speculate that technology has the potential to (and already does) work with both the drive system and the affiliative system, but that it has (like our society) rather disproportionately targeted the drive and excitement system. This is unsurprising because digital technologies were largely conceived as productivity and work tools to begin with, and perhaps if they had stayed in the workplace, the disproportion wouldn't be so inappropriate. But now digital technologies take part in everything, from our intimate relationships to the ways we care for ourselves and others, so what we might come to refer to as "affiliative design," or design to support the affiliative system and its associated positive emotions, seems far more appropriate for many of these contexts.

For example, no one would have considered inviting a mainframe into bed in the 1960s, but today people can use computers to track their sex lives. Apps such as Spreadsheets and Sexulator show just how far our productivity and achievement mindset goes. However, one comment from a reader of a women's magazine is revealing. Her favorite aspect of a sex-tracking app wasn't the tracking at all, but rather the new opportunity for expressing love playfully: "My favorite part: being able to leave messages for my husband. I'd say, 'You were really hot last night!' or 'That was fabulous! Hope we can do it again soon!' He was pleasantly surprised, and it was fun to have our little secret."[3] Features that supported the affiliative rather than achievement-driven positive emotions added value to her experience.

What's potentially problematic is that it's easy to see how tracking things such as intercourse that are intended to be spontaneous and genuine expressions of human connection could lead to dissatisfaction, comparison, guilt, self-criticism, and so on. Even if this drive-based achievement focus leads to positive emotions such as pride, how might it shift us away from the affiliative motivations and emotions more capable of enriching our relationships? And how far do we go with such tracking? Would we encourage children to track how many hugs they get from their parents? What's important is to acknowledge that how we frame an activity through the design of technology, what positive emotions we design to support, what rewards and features we focus on, can change how users approach and feel about the activity itself—as well as how much they can benefit from the technology psychologically. Design is not a one-way street. Using a technology will have consequences on how we view and engage with the activity it is meant to enhance or monitor. Tracking intimacy will affect how we view and engage in intimacy.

Therefore, just as extrinsic motivation can undermine intrinsic motivation (which we discuss in the next chapter), we speculate that activating our drive system has the capacity to undermine the activation of our affiliative system, which might have potentially negative consequences in certain contexts. But the flip side is that embracing design for the affiliative system could prove to be an antidote to the negative side effects of an overfocus on the drive system. For example, some early research suggests that design for affiliative experience might help curb addiction. Murat Iskender and Ahmet Akin (2011) found a link between self-compassion and a reduced likelihood of Internet addiction. Such a correlation holds incredible promise for future designs that intrinsically guard against imbalance by including support for affiliative emotions.

Therefore, let us innovate creative ways to develop digital opportunities for sharing kindness, expressing compassion, feeling admiration, relishing gratitude, enjoying contentment—rather than simply always falling habitually back on the promotion of wanting and striving. Our wellbeing, our families, our communities, and ecological sustainability may depend on it.

Strategies for Cultivating Positive Emotions

Although we have tried to make the point that wellbeing goes beyond positive emotions and that our assessment of them is more complex than we may realize, positive emotions nevertheless remain both critical to theories of wellbeing and an obvious target for technological design. So how does one go about increasing positive emotions through technological design without falling pray to oversimplistic attempts? "Be angry" is poor direction for a stage actor, and "don't be so sad, cheer up!" is fairly useless advice from a well-meaning friend. Technology designers should be wary to avoid the similar trap of trying to tell people what to feel.

A skilled director might tell her actor to bang his fist (a concrete instruction that both shows and elicits anger), a helpful friend might suggest we go for a walk together, and a therapist might encourage us to identify unhelpful thinking patterns, keep a journal, or improve our sleeping habits. We suspect that it is also to more concrete behavior, activities, and conditions (those that improve wellbeing according to research) that we as designers of technology are likely to turn for actionable design strategies for positive computing.

As a starting point, we will need a scope for the rather general term *positive emotions*. Fredrickson (2001) lists 10 emotions that she considers significant to wellbeing and phenomenologically distinct: joy, gratitude, serenity, interest, hope, pride, amusement, inspiration, awe, and love. Fredrickson also points out that there are two ways of increasing your overall experience of positive emotion: by reducing negative emotions or by increasing positive ones.

Positive-psychology interventions—that is, "treatment methods or intentional activities that aim to cultivate positive feelings, behaviors, or cognitions" (Sin & Lyubomirsky, 2009)—represent the best collection of research-based strategies for promoting positive emotion. A meta-analysis by Nancy Sin and Sonja Lyubomirsky (2009) contained 51 randomized control trial studies, encompassing 4,235 participants. The r effect sizes ranged from –0.31 (studies where unexpectedly the control group had a better outcome than the treatment) to 0.84 with a mean of 0.29. Most studies compared a pretest with 8-week and 12-week follow-ups. The interventions included a whole spectrum of activities: mindfulness, gratitude, optimism, goal setting, and positive writing. Although the study does not report if any of the interventions were delivered with technology, there are online examples for many of these interventions.

The strategies noted here were purposely built for positive psychology, but we can also look to research on the psychological impact of commercial technologies. For example, recent randomized control studies on the use of casual games at East Carolina University (Russoniello, O'Brien, & Parks, 2009) showed that they generated positive emotions to the extent that they were effective in significantly reducing levels of stress, anxiety, and depression. Furthermore, a recent review of videogames from the perspective of their impact on young people's wellbeing (Johnson, Jones, Scholes, & Colder Carras, 2013) concluded, "Research suggests that videogames contribute to young people's emotional, social and psychological wellbeing. Specially, videogames have been shown to positively influence young people's emotional state, self-esteem, optimism, vitality, resilience, engagement, relationships, sense of competence, self-acceptance and social connections and functioning." (For more on the astounding potential of videogames to support wellbeing, see Jane McGonigal's sidebar in this chapter.)

Similarly, research on social networks, which has turned up both positive and negative impacts on wellbeing, is invaluable to informing future designs and redesigns of this genre of software. One study (Mauri, Cipresso, Balgera, Villamira, & Riva, 2011) collected physiological data from 30 subjects and found evidence that looking at the Facebook homepage triggered physiological patterns resembling positive emotions (at least more so than doing math problems or looking at panoramas). A different study (Chou & Edge, 2012) suggested an apparently contradictory effect: "Those who have used Facebook longer agreed more that others were happier, and agreed less that life is fair, and those spending more time on Facebook each week agreed more that others were happier and had better lives." Without a doubt, we need more research that helps us understand the complex and multifaceted impact of social networking on our psychological lives.

As important as identifying what works is taking into consideration the specific parameters and limitations that make it work. Every intervention has "optimum settings." For example, for the "random acts of kindness" intervention, it is more effective if the activity is carried out not as daily routine, but rather intermittently. In a similar way, the capacity for digital games to nurture positive emotion will be impacted by the schedule and number of hours they are used. A recent review of videogames for wellbeing (Johnson, Jones, Scholes, & Colder Carras, 2013) noted that "how young people play as well as whom they play with may be more important in terms of wellbeing than what they play."

Finally, it's critical to keep in mind that when we speak of technology, we speak of highly complex systems that intermingle with multiple aspects of our lives and minds in many different ways. One system (e.g., a social network or a multiplayer game) is bound to elicit a large combination of both positive and negative emotions relating to a large combination of design decisions. Rather than seeking oversimplistic equations (such as social networks = good, or videogames = bad), we need to anticipate that the impacts will always be mixed and the process of untangling the complexity will always be ongoing.

Design Implications

If someone thinks an experience they had was better than it really was, is that a bad thing? The implications of the colonoscopy study suggested to doctors that designing for a better remembered experience should take precedence over designing for a better moment-by-moment one. This seemed reasonable because, in the medical case, designing for better memory leads to greater wellbeing in the long term and greater compliance with repeat visits.

Creating a memory-erasing device would probably be taking this logic too far, but in most cases of technology design, designing to optimize *remembered* experience seems to make a great deal of sense. Indeed, Don Norman (2009) argues that seeking a good remembered experience is more important than seeking a perfect moment-by-moment user experience and points to Disneyland, iPods, and trips to Thailand as perfect examples. Neither endless lines nor poor usability nor squat toilets can keep these experiences from turning rosy in retrospect. In his article for the journal *Interactions of the ACM*, "Memory Is More Important Than Actuality" (2009), he points out that "[when we remember] events, some things fade from the mind faster than others. Details fall away faster than higher-level constructs. Emotions fade faster than cognitions." Our reflective ideas about an experience will last longer than the fleeting frustrations that formed part of that experience at the time. "So, make sure the beginning and the end are wonderful," he concludes. "Make sure there are reminders of the good parts of the experience: Photographs, mementos, trinkets. Make sure the experience delights, whether it be the simple unfolding of a car's cup holder or the band serenading departing cruise-ship customers. Accentuate the positive and it will overwhelm the negative."

Entries and Exits
Bringing together the peak–end rule, Norman's notes on serenades, and what we know about the importance of first impressions, we can only conclude that designers, like good playwrights, need to attend to their entrances, their climaxes, and their exits.

In terms of entries to experience, there's already plenty of design research and experience to show how powerful first impressions can be and therefore how much care is put into homepages and openers, but do we carefully consider the way users exit our experiences? When someone leaves our website, do we send them off with a release of balloons and applause, a fond farewell, or a carefully unfolding denouement? Probably not. To better conceptualize our digital exits is to, as Tom Stoppard's Player suggests, "think of every exit as being an entrance somewhere else."[4] This not only better serves the peak–end rule but takes into consideration the reality of the user's continuous experience.

In order to embrace our exits, we need to better understand where our users go after they interact with our product. We might ask ourselves how a memory of the experience can be retained and shared with others. Many digital interactions designed for children already allow kids to print their creations, or show evidence of their achievements in the form of, say, certificates or trading cards. Amazon's Kindle app detects when you have reached the last page of the book you have been reading and uses this moment of near closure to offer you exit options. You can rate the book with one tap, share that you've finished it on a social media site, or choose from a list of books by the same author or on the same topic. Of course, this design decision is about generating business, not wellbeing, but it does show how we can begin to do more with our exits.

It's worth noting that acknowledging the importance of remembered experience should not be mistaken as disregard for the importance of present-moment attention. We're looking at how remembered experience can impact design decisions, which does not mean we discourage the user from a mindful experience—the two are not in opposition. And, of course, mindfulness is not an approach to the design of technology, but a state of mind and a practice that we look closely at in chapter 10. Of course, we don't argue that the remembered experience is the *only* important thing (otherwise we'd be focusing on developing that memory-erasing stick). Remembered experience is simply an aspect of our wellbeing worth taking into consideration, and doing so is a useful antidote to seeking a flawless user experience, which is both impossible to achieve and sometimes overvalued in terms of longer-term benefits.

Kahneman's (2000) method for measuring what he refers to as "total utility" aggregates moment-by-moment assessments, conducted over a period of time, and he deliberately includes those moments spent reflecting about a past experience as part of the measure of that past experience. Indeed, one's method of measuring wellbeing will have an impact on how one designs for it. If positive recollection is of importance to your wellbeing measure, you will probably choose to invest in supporting recollection in the design of a digital experience.

Still, simply attending to peaks and exits is unlikely to be the only answer to cultivating positive emotion for wellbeing. Designers have been reveling in the hedonic quality of digital experience in its own right for years, both because it's fun and because it's smart business.

Design for Emotion
Designers love to think and talk about how their designs might elicit delight, fun, playfulness, happiness, calm, excitement, pride, or even awe. After all, we're a playful bunch who want to make the world more enjoyable. It is arguably because Apple has triggered so many of these emotions so consistently that its series of mobile devices has become so wildly popular. And as Norman (2009) points out, it is also for this very reason that its many usability problems get consistently overlooked by loyal users. The frustration that might lead to rage with other computer technologies has been sealed out for Apple devices by a thick resilient buffer of positive emotion.

Less obvious perhaps are the many ways designing for positive emotions can have useful cognitive consequences. How often do we design for positive emotion specifically to affect other task goals such as productivity, learning, and problem solving? Many research studies have shown in contexts as varied as boardrooms and doctor's offices that positive emotion increases creativity and problem solving. In his book *Emotional Design* (2005), Norman points out that "emotions, we now know, change the way the human mind solves problems—the emotional system changes how the cognitive system operates. ... Positive emotions are critical to learning, curiosity and creative thought. ... [B]eing happy broadens the thought processes and facilitates creative thinking."

Norman's framework for understanding the different levels at which we react emotionally to design divides our experience into the visceral, behavioral, and reflective. The visceral includes aesthetics and first impressions and engages our senses. Visceral design alone can make us feel excited, relaxed, anxious, or wary. At the behavioral level, it is what we do, how we move, the activity we engage in, how we control and interact with a product, and how it gives us feedback that can be satisfying, joyful, or aggravating. Finally, the reflective level engages our intellect, our values, our opinions, and our judgments. At this level, we consider how a product reflects our values, supports our goals, adds to our identity or status, how it contributes to a more sustainable or more just world, how it represents family or connectedness—all of these webs of concepts that we build about a product or experience come with emotions such as pride, confidence, motivation, affection, and so on, and, as mentioned earlier, it is the reflective level that is most persistent.

Marc Hassenzahl and Andreas Beu (2001) have emphasized that we ought to understand the nature of hedonic quality as a software requirement, citing its importance to product satisfaction, credibility, and usability as well as the potential effects it can have on the quality of the work being done with that software: "in certain work positions (those requiring 'emotion work,' such as a call center agent or hotel receptionist), enjoyment might have an important effect on work quality instead of solely serving a therapeutic purpose." And they propose a number of techniques, including a repertory grid and interview techniques, for measuring the hedonic quality of software.

Clearly, emotion occurs at many levels of a user experience, and the more we can understand about these levels, the better we can apply this knowledge in the service of wellbeing. We have now seen the potential for increases in positive emotions to trigger an upward spiral of flourishing and have noted that affiliative positive emotions are essential to creating the balance that defines that flourishing.

But positive emotions alone are not enough to complete the positive-computing picture. Positive emotions co-occur with other factors such as mindfulness, compassion, motivation, engagement, and flow, which is where we head next. In the following chapter, we look at when digital experiences motivate people to connect, take action, and change, which, when considered en masse, can have strikingly powerful implications.

Expert Perspectives—Technology for Positive Emotions

Fun and Pleasure in Computing Systems

Don Norman, Nielsen Norman Group

The design of human–computer systems used to focus upon the negative, the breakdowns that confused and confounded people. Now it is time to move to the next level, to focus upon the positive, systems that are enjoyable and pleasurable. We need systems that delight as well as inform, systems that create pleasure along with having a useful function. We need systems that are resilient, that promote control, understanding, and sometimes just plain pleasure. The design field has responded by examining the role of emotions and pleasure in design. We need to move these findings into mainstream computing.

Modern gesture devices provide physical pleasure. It's delightful to "toss" a file to another person or to slide photographs across a virtual tabletop, rotating, stretching, or shrinking them. I am a fan of the lovely "bounce" that a list of scrolling items does when it reaches the end, and then I enjoy trying to pull down the top item from the top of the screen, seeing how far I can get it to move until it resists and springs back to its proper location. Do all these gestures, movements, stretching, and bouncing provide function? No, but who cares? They provide pleasure.

Beautiful aesthetics and creative fun are equally important. Engineers and businesspeople are apt to object: "What has that got to do with getting the job done?" they will ask. But note: many master chefs in the world's best

(continued)

> restaurants spend as much time on the presentation of the food as on the ingredients and cooking. "Attractive things work better," I once wrote, and they do, if only because a positive mood quickly dismisses as trivial and irrelevant minor difficulties that can trigger a state of irritability when in a negative mood. I watch people struggle to understand how to do things on their smart phones, only to finish and tell me, smiling, "That was easy." The moral is simple: we must consider the presentation of our computer applications.
>
> Lots of psychological research supports the observation that we remember best and weigh most highly the ending of an experience. (Next highest is the start—of least importance is the middle.) End well, and in one's memory the total experience was great. Difficulties are often unavoidable in the performance of complex tasks, but design for fun and pleasure with a positive and uplifting ending, and all will be forgiven. Memory triumphs reality. After all, an experience exists only at its moment of occurrence: the memory of the experience lives on long afterward. Design for the memory. Positive computing? It is about time.

(continued)

> **Let the Positive Games Begin**
>
>
>
> Jane McGonigal, Institute for the Future
>
> It's time to take games seriously as a platform for increasing global wellbeing. More than one billion people on this planet play videogames regularly, an average of one hour per person each day. Just imagine: What if we could convince a whopping *one billion people* to spend an *entire hour* every single day of their lives investing in their own wellbeing—by provoking powerful positive emotions, stoking their sense of engagement with difficult challenges, strengthening their social relationships, building up self-efficacy, and even connecting them to a greater purpose?
>
> Game designers already possess this power—and, fortunately, many are starting to use it consciously for good. Increasingly, games are designed explicitly to have a positive impact on players' real lives. Cooperative games such as *Minecraft*, where players work together instead of against each other, is on the rise. Co-op now outranks competitive gameplay in both hours played and dollars spent, leading to stronger social bonds between players. Meaningful gameplay is also becoming more common, with games for change that tackle real-world problems—such as the Facebook game *Half the Sky*, which empowers players to make donations to schools for girls in developing regions, or *FoldIt*, which helps players team up with scientists to research new cures for cancer.

(continued)

> Casual games such as *Bejeweled* and *Peggle* have been successfully tested in randomized controlled studies as a nonpharmaceutical treatment for anxiety and depression. And it's not just mental pain that can be treated—the 3D immersive environment *Snow World* has proven more effective than morphine at treating extreme physical pain in burn victims. Meanwhile, research on the effects of playing a game with an avatar different from yourself in gender, age, weight, or ethnicity has led to virtual environments designed specifically to increase player compassion for and decrease bias against women, the elderly, the obese, and people of a different ethnic background.
>
> More positive emotions on a daily basis triggers an upward spiral of happiness, health, and success, meaning a game you enjoy can do a world of good. Even just a few moments of feeling authentically happy can get players' positive-to-negative emotions above the magic three-to-one ratio recommended by scientists such as Dr. Barbara Fredrickson. That's why many game design curricula at universities and industry conferences now emphasize the wide range of positive emotions that games can create, from curiosity and surprise to awe and wonder.
>
> These innovations are just the beginning. We have only just started to realize the potential of games to increase happiness, improve life satisfaction, and bring more meaning and purpose to players' real lives. The next decade will be full of playful experiments that build on positive-psychology research—so let the positive games begin!
>
> • *Resources*: Games for Change, gamesforchange.org; Games for Health, gamesforhealth.org.
> • My blog *Show Me the Science* compiles research on real-life positive impacts of games: Showmethescience.com (all of the research mentioned here is documented at this site).

Notes

1. From *An introduction to the principles of morals and legislation* (1789, 1823).

2. Pollyanna is the main character of the 1913 children's novel *Pollyana* by Eleanor H. Porter. Pollyana is particularly optimistic, always looking for the positive side of things—some would say to an excessively unrealistic extent. Sociopaths are unable to experience certain negative emotions such as guilt and remorse or to empathize with others emotionally.

3. See "Track your sex life frequency," in "17 hot new things to try with your guy," *Redbook* online, redbookmag.com/love-sex/advice/spice-up-sex-life-18#slide-10, accessed February 14, 2014.

4. From *Rosenkrantz and Guildenstern are dead* (New York: Grove Press, 1967).

References

Burkeman, O. (2013). *The antidote: Happiness for people who can't stand positive thinking*. New York: Faber and Faber.

Chou, H.-T. G., & Edge, N. (2012). "They are happier and having better lives than I am": The impact of using Facebook on perceptions of others' lives. *Cyberpsychology, Behavior, and Social Networking, 15*(2), 117–121.

Diener, E., & Biswas-Diener, R. (2008). *Happiness: Unlocking the mysteries of psychological wealth*. Oxford: Blackwell Publishing.

Fredrickson, B. L. (2001). The role of positive emotions in positive psychology: The broaden-and-build theory of positive emotions. *American Psychologist, 56*(3), 218–226.

Fredrickson, B. L., & Losada, M. F. (2005). Positive affect and the complex dynamics of human flourishing. *American Psychologist, 60*(7), 678–686.

Gilbert, P., & Choden (2013). *Mindful compassion* (p. 384). London: Constable & Robinson.

Hassenzahl, M., & Beu, A. (2001). Engineering joy. *IEEE Software, 18*(February), 70–76.

Hochschild, A. R. (2003). *The managed heart: Commercialization of human feeling*. Berkeley: University of California Press.

Isen, A. M. (1990). The influence of positive and negative affect on cognitive organization: Some implications for development. In N. L. Stein, B. Leventhal, & T. Trabasso (Eds.), *Psychological and biological approaches to emotion* (pp. 75–94). Hillsdale, NJ: Lawrence Erlbaum Associates.

Iskender, M., & Akin, A. (2011). Self-compassion and Internet addiction. *TOJET*, *10*(3), 215–221.

Johnson, D., Jones, C., Scholes, L., & Colder Carras, M. (2013). *Videogames and wellbeing: A comprehensive review*. Melbourne, Australia: Young and Well Cooperative Research Centre.

Kahneman, D. (2000). Experienced utility and objective happiness. In D. Kahneman & A. Tversky (Eds.), *Choices, values and frames* (pp. 673–692). New York: Cambridge University Press and the Russell Sage Foundation.

Kahneman, D. (2011). *Thinking, fast and slow*. New York: Farrar, Straus and Giroux.

Lyubomirsky, S., King, L., & Diener, E. (2005). The benefits of frequent positive affect: Does happiness lead to success? *Psychological Bulletin*, *131*(6), 803–855.

Mauri, M., Cipresso, P., Balgera, A., Villamira, M., & Riva, G. (2011). Why is Facebook so successful? Psychophysiological measures describe a core flow state while using Facebook. *Cyberpsychology, Behavior, and Social Networking*, *14*(12), 723–731.

Norman, D. A. (2005). *Emotional design: Why we love (or hate) everyday things*. New York: Basic Books.

Norman, D. A. (2009). Memory is more important than actuality. *Interactions of the ACM*, *16*(2), 24–26.

Redelmeier, D. A., & Kahneman, D. (1996). Patients' memories of painful medical treatments: Real-time and retrospective evaluations of two minimally invasive procedures. *Pain*, *116*, 3–8.

Russoniello, C. V, O'Brien, K., & Parks, J. M. (2009). The effectiveness of casual video games in improving mood and decreasing stress. *Journal of CyberTherapy and Rehabilitation*, *2*(1), 53–66.

Sin, N. L., & Lyubomirsky, S. (2009). Enhancing well-being and alleviating depressive symptoms with positive psychology interventions. *Journal of Clinical Psychology*, *65*(5), 467–487.

Story, L. (2007). Anywhere the eye can see, it's likely to see an ad. *New York Times*, January 15.

7   Motivation, Engagement, and Flow

I can hear the threatening moans of the undead gaining on me from behind. Picking up the pace, I break into a run, and my heart quickens. I round a sharp corner, cut through the park, and finally welcome the reassuring news through my earbuds: "Zombies evaded." I pick up some virtual medical supplies and head for home.

Jogging for your life in the midst of a zombie apocalypse is just one of the many ingenious ways designers have conceived to get people motivated in the modern world.[1] The reality is, I'm not the Nike+ type (Dorian, here). I don't feel like an amazing athlete, not even with a wristband or strategically triggered applause. But immerse me with agency in the unfolding of a satirical suspense narrative, and I've managed a heart-pumping run through the neighborhood—all before breakfast.

Motivation and wellbeing intermingle in sophisticated ways. Not only is motivation fundamental to taking any kind of positive action, but the absence of it is a hallmark of depression. Clearly, a life rich in motivation is more rewarding than life without.

Motivation is a trigger to act, and when that activity is sustained by an ongoing urge to carry on, we are *engaged*. We might be engaged in writing an article, playing Frisbee, or laying bricks to build a house for someone homeless. Our level of engagement may be barely enough to keep us going, or it might be all-encompassing, sweeping us into a state of *flow* (that place of optimal engagement famously described by Mihaly Csikszentmihalyi.)

It's hard to imagine a technology designer who wouldn't be aiming to motivate users in some way, be it to download, upload, collaborate, contribute, click through, or learn more. As such, there are many resources with advice on how to do so more effectively through design. This chapter distinguishes itself by taking the less considered angle of motivation as a contributor to psychological wellbeing. We look at the interrelated notions of motivation, engagement, and flow. We consider key motivation theories, giving special attention to those that provide an explicit bridge between motivation and wellbeing. We then turn to current technologies to seek out hints of how motivation theory might be applied to the design of future things to increase users' psychological wellbeing.

Motivation

The Pleasure Principle

At the most primitive level, motivation can ultimately be viewed as the desire to seek pleasure and avoid pain. With origins stretching as far back as ancient Greece at least, our understanding of motivation has at various points been contested, reinforced, and expanded by psychologists, economists, and philosophers who have integrated concepts as diverse as altruism, primal urges, autonomy, cognitive dissonance, and interconnectedness into the mix.

Motivational theories, too numerous to cover here, have generally focused on either social underpinnings, psychological drivers (e.g., cognitive), and biological factors. We look at several of the major theories and issues and begin with a point of contention impossible to omit from any discussion of motivation.

Intrinsic and Extrinsic Motivation—a Sibling Rivalry

If I engage in an activity because it's fun, I am said to be intrinsically motivated. In a sense, the activity is its own reward. If I engage because I fear the stick or crave the carrot, I am said to be extrinsically motivated. The carrot represents a reward separate to the task (e.g., money, points, or approval), and the stick is, of course, a punishment (e.g., exclusion, demotion, or imprisonment), each of which resides outside myself and is controlled by some mechanism external to me (a parent, a boss, or a judge). As Richard Ryan and Edward Deci (2000) put it, "The most basic distinction is between intrinsic motivation, which refers to doing something because it is inherently interesting or enjoyable, and extrinsic motivation, which refers to doing something because it leads to a separable outcome."

Our environment is replete with examples of extrinsic motivators, and most adults now living were educated by a system that relied on it almost exclusively. Nowadays, the mere mention of a stick is a bit cringeworthy and suggests the controversies that have arisen over what motivators are effective, desirable, or just. Although many psychologists and educators have all but abandoned the stick, researchers such as David Greene, Mark

Lepper, and Edward Deci have warned that the carrot can be similarly destructive.

In an often cited seminal study (Greene & Lepper, 1974), preschoolers offered a reward to do something intrinsically motivating (draw pictures) lost their intrinsic motivation and drew *less* than children asked to do it without a reward. This is just one in a slew of similar studies that have exposed the potential for extrinsic motivators to undermine intrinsic motivation, and growth in this area of research has led to a number of modern revisions of how we should structure our workplaces and societies.

Offering *contingent* rewards ("if you do this, then you get this") can turn something enjoyable into work, a shift that, over time, degrades intrinsic motivation and may condition us always to need rewards to be motivated. Because intrinsic motivation is associated with quality learning, felt competence, persistence, creativity, positive coping, and wellbeing, then sabotaging it is counterproductive and, some would suggest, a contributor to society-wide problems.

But, obviously, we can't always be intrinsically motivated to do all that needs to be done in a day, and this is where extrinsic motivators become important. Indeed, Ryan and Deci (2000) argue that certain kinds of extrinsic motivation share many of the benefits of its intrinsic sibling and that the important difference lies in autonomy. They provide a model that may prove invaluable to work in positive computing because it separates extrinsic motivation into four categories, each with a "perceived locus of causality" that is more or less externally derived:

- **External regulation**, which is entirely "external" (e.g., compliance).
- **Interjection**, which is "somewhat external" (e.g., seeking approval).
- **Identification**, which is "somewhat internal" (e.g., activity is valued).
- **Integration**, which, like intrinsic motivation, is "internal." In this case, external regulators are assimilated to the self.

For example, a student memorizing a list of pharmaceuticals because she's desperate to graduate med school and become a doctor is acting in a way that is self-determined, even though the memorizing itself is not intrinsically fun. She has *identified* the task with a life goal and the causality can therefore be considered "somewhat external." Someone who volunteers his time to stuff envelopes for Amnesty, isn't doing it because stuffing envelopes is a blast, but because he is motivated by a compassionate desire to help others and seek justice. The motivation is extrinsic to the task, but highly self-determined and therefore beneficial and rewarding. The task aligns with his core values, and he may have fully *integrated* this type of activity with himself.

Ryan and Deci's (2000) review of a number of studies reveals that extrinsic motivation that is more *internal* (and therefore more *autonomous* or self-determined) is associated with greater engagement, better performance, higher-quality learning, and *greater psychological wellbeing* (many of the same benefits attributed to intrinsic motivation).

Social influence such as reciprocity, gratitude, positive self-image, and career goals are all commonly observed motivators for people who engage with social media, all of which can be mapped to Ryan and Deci's taxonomy of human motivation. In thinking about design for wellbeing, it is useful to know that the range of motivators that are more *internally* derived will better support wellbeing. Thus, aiming to support more internally derived forms of motivation rather than relying too heavily on simple contingent rewards (or the simplistic imposition of shallow game mechanics) is a worthwhile pursuit in the context of positive computing.

The most obvious deployment of every type of intrinsic and extrinsic motivation in technology design today is mastered in games and seen in the application of game mechanics to nongames, also known as "gamification." According to information technology and research advisory company Gartner (2011), by the time you read this, "more than 50 percent of organizations … will [have] gamif[ied] [innovation] processes." Gamification rewards can be used in many ways, including those that undermine intrinsic motivation, but also in effective ways that add motivational layers of enjoyment to inherently unpleasant tasks or as feedback to reflect growing competence. We look at some examples later in this chapter.

Motivation That Is Intrinsic to Being Human

In the case of the volunteer and the medical student mentioned earlier, it may be helpful to look at their goals as stemming from innate human needs, such as purpose, connectedness, competence, and self-actualization (depending on the theory you employ). Although much motivation is contingent on individual interests, some drivers are considered universal to being human. Most obviously there are the physiological drives that urge us to satisfy hunger, protect ourselves from the elements, and procreate. Moreover, many games are built on the natural human motivation to seek patterns in visual information, collect things, connect with others, and resolve cognitive dissonance (sometimes manifest as mysteries, puzzles, or conflicting information). Some of the key theories describing such innate human motivators are described in the next subsection.

Drives, Needs, and Desires
At the foundation of modern motivational theory sits Abraham Maslow's (1943) hierarchy of human needs. According to this influential theory, we are driven by five levels of needs: "It is quite true that man lives by bread alone—when there is no bread. But what happens to man's desires when there is plenty of bread and when his belly is chronically filled?" Maslow goes on to answer this question with a list of need categories, each prerequisite to the next, suggesting that a new category is "unlocked" only once the previous category has been reasonably satisfied.

1. **Physiological needs**, such as food, air, and sleep are primary. Only when these needs are met (and, Maslow argues, they generally are outside of emergency situations) can humans move on to other needs.
2. **Safety** refers to our need for security and stability as well as to safety from physical danger. Insurance plans, career decisions, savings accounts, burglar alarms, and deep freezers can be looked on as ways in which we are motivated by our need to feel relatively protected from harm and loss. Our desire to accrue, collect, and build things, be they real supplies or achievements and virtual collectables in a game, might be linked to this underlying need for stability and safety nets.
3. **Love** follows once the first two need categories are fairly well satisfied, and it includes belongingness and the giving and receiving of affection. Various types of digital environments allow us to develop a sense of belongingness to a group, connect with people whom we love or may come to love, and share in affection. (Sending an intimate text message, "poking" someone, or using the kisses emoticon in a chat box can be interpreted as virtual signs of affection (or "affiliative design," as mentioned in chapter 6).
4. **Esteem** or the high evaluation of oneself by oneself and by others is a recognizable need. Maslow points specifically to esteem-related desires for strength, achievement, adequacy, confidence, reputation, independence, and freedom. (We discuss self-awareness, self-esteem, and self-compassion in detail in chapter 8.) Much of the way we measure our sense of worth in the modern world is mediated by technology. Whether it's in endorsements, profiles, "likes," or eXperience points, technology has an undeniable impact on our capacity to feel and extend esteem.

5. **Self-actualization** refers to a person's tendency to reach his potential, to achieve the most he can become, and to feel fulfilled. As Maslow describes it, "A musician must make music, an artist must paint, a poet must write, if he is to be ultimately happy. What a man can be, he must be. ... This tendency might be phrased as the desire to become more and more what one is, to become everything that one is capable of becoming." New models of open education such as open content and massive open online courses provide opportunities for people to explore creative potentials or pursue mastery. Ideally, positive computing will come to increasingly support people in building human psychological potentials such as mindful awareness, compassionate action, and emotional intelligence.

In the search for fundamental human needs, others have followed Maslow. Steven Reiss (2004) has proposed a set of 16 basic desires, most of which can be filed into Maslow's hierarchy, with the exception, perhaps, of idealism, power, and vengeance. More recent theories on human motivation have given concepts related to autonomy and competence a more central role, such as Ryan and Deci's SDT mentioned in chapter 2. Also among the most notable is the work of Carol Dweck.

"I Think I Am" versus "I Think I Can"—Fixed and Growth Mindsets

In her early work, Stanford University psychologist Carol Dweck (Dweck & Leggett, 1988; Dweck, 2006) identified two types of behavioral responses: a "helpless" pattern in which people tend to avoid challenges, view obstacles negatively, and reduce performance; and a "mastery-oriented" pattern observed among people who seek challenges and who are persistent in the face of obstacles. In her book *Mindset* (2006), she describes these patterns of behavior as being related to "fixed" and "growth" mindsets respectively.

Simply stated, those with fixed mindsets believe their abilities arise from innate capabilities and intelligence endowed at birth that cannot be changed. In contrast, those with a growth mindset believe their abilities are developed over time and can be enhanced, a view that is in far greater alignment with recent discoveries on neuroplasticity and epigenetics. It turns out that these two subtle variations in how we view ourselves lead to striking differences in behavior and wellbeing.

According to Dweck's research, those harboring a growth mindset are significantly better at identifying their strengths and weaknesses, and when faced with a setback, they tend to look for learning opportunities. In contrast, those with a fixed mindset are more focused on judgments.

When a fixed-mindset person is confronted with a setback, her tendency is to judge herself, as in "I am a failure." Even when outcomes are positive, there is a tendency to compare herself to others: "I am better than the others." Research has shown that the fixed mindset is more likely to lead to unhappiness.

Cognitive-behavioral therapy helps people detect helpless patterns and is one of the most successful modern strategies for helping people with depression. The work of Dweck and others such as Heidi Grant, a motivational psychologist at Columbia University's Motivation Science Center, have been influential to practices in education, management, and personal development. Some applications of this work include advice to parents and caregivers to praise children for effort and improvement rather than innate intelligence ("Good job, you've really improved" or "You worked hard for that" rather than "Good job, you're so smart" or "I bet you are the smartest in your class").

These findings can also inform the design of applications that help people set and follow up with goals. Goal-setting tools span a broad spectrum, including those that draw explicitly on psychological theory as well as those that focus more on technical sophistication. What Dweck's research shows us with regard to wellbeing is that it's not just the goals you set that are important, but how you think about them and about yourself in relation to them. Therefore, goal-setting tools can impact wellbeing not only by supporting wellness-related goals (I vow to run more or eat better), but also in terms of how they support thinking around goal setting. Trash talk, leaderboards, task breakdown, and deadlines are all approaches to motivating someone toward reaching a goal, and these strategies may or may not have positive impacts on wellbeing depending on the context. Goal setting has been shown to be highly effective in many contexts, but it is not entirely free from caveats when it comes to wellbeing.

Goal Setting—Implications for Wellbeing

Goal-setting theory has provided a framework for investigation into how and under what circumstances defining goals influences things such as performance, self-efficacy, and satisfaction, particularly within organizational and educational settings. Explicit goals are used as motivators, and researchers such as Edwin Locke and Gary Latham have consistently found that setting specific, difficult goals leads to higher performance by individuals and teams. As such, goals are often used, especially in workplaces, to boost productivity and achievement. However, Lock and Latham's research also shows that the effectiveness of goal setting is dependent on contextual factors such as task difficulty, competence, framing, and self-efficacy. They discuss some of these moderators of goal effects in their essay "New Directions in Goal-Setting Theory" (Locke & Latham, 2006).

Others have examined the effects of goal setting, goal attainment, and disappointment in the area of personal development and mental health. At a broad level, research has found that "individuals valuing relationship goals above achievement oriented goals have been found to have a greater sense of wellbeing than individuals placing achievement goals above relationship goals" (Street, 2002). Helen Street also reviews how goals can either worsen or improve depression and how variables such as goal definition, content, type, and framing play a part. She also explores those situations in which people relentlessly pursue goals regardless of personal cost.

Some have questioned what they consider to be the overuse of goal setting, particularly in American culture. D. C. Kayes (2004) used the 1996 Mount Everest disaster (in which six expert climbers climbed to their deaths despite all safeguards and information urging them to return) as an example of when relentless adherence to an original goal can become irrational and cause team breakdown. "In the face of an environment that requires learning, short-term project teams may encounter the limits of the positive effects of goal-setting."

More research is needed before we can fully understand the effects of goal setting on practices such as mindfulness and positivity. Mindfulness is frequently described as a method for letting go of plans for the future and nonjudgmentally settling into the realization that "just being is enough." Goal setting seems decidedly antithetical in this context. Yet the current standard for most wellness technology is to apply tracking features that can encourage both goal setting and comparison to others ("I will practice mindfulness every day this month" or "I failed to be grateful 10 times this week" or even "I can't believe my brother beat me on meditation today"). For certain types of positive-computing technologies, it may be the *relinquishing of goal setting* that requires support, at least for some user groups and particularly because many of us currently live in otherwise heavily goal-driven societies.

Clearly, technology designed to promote motivation and wellbeing through goal setting must keep in mind the need for balance and be skeptical of overly simplistic views of goal-setting psychology. As always, a technology team's greatest safeguard is collaboration with mental health professionals.

There is still much investigation to be done around the effects of various types of tracking, goal setting, self-evaluation, and game mechanics with regard to their roles in appropriate balance—for example, balance between positive thinking and heartfelt authenticity, between directedness and present-moment attention, between drive and calm, desire and contentment, dissatisfaction and acceptance, and other balance relationships related to goals that impact psychological wealth and wellbeing.

Social Motivators

Another motivator intrinsic to being human lies in our tendency to be influenced by the actions and opinions of others. Social psychologist Erving Goffman (1959) proposed that much of human behavior is motivated by how we would like others to see us, and his theory provides a way of conceiving the public versus private lives we keep online. Goffman's interactionist theory was based on his "dramaturgical approach," in which behavior is seen as a series of minidramas in which *individuals* perform in front of an *audience*, and both are *participants* in a *performance*. The performance has a *front stage* on which the audience sees the individual and a *backstage* representing where the individual is when alone. Some have described Goffman's work as "social phenomenology" (Miles & Huberman, 1994).

Sunny Consolvo and Katherine Everitt (2006) propose a set of guidelines based on Goffman's and Leon Festinger's (1957) theories for designing systems that encourage physical activity: (1) give proper credit; (2) provide history, current status, and performance measures; (3) support social influence (i.e., use social pressures and support); and (4) consider practical constraints. These guidelines have since contributed to successful systems (Consolvo, 2009a) as well as to more generic and theory-driven guidelines for behavior change (Consolvo, 2009b).

Consolvo (2009b) argues that technologies that support behavior change should support *impression management*, the individual's movement between Goffman's "front-stage" and "back-stage" behaviors. She urges that these technologies need to allow users to manage backstage access. For example, if a user wants to misrepresent an event or conceal an action, the system should support this type of behavior. (Not everyone wants his or her personal foibles or failed exercise routines made public.) This requirement offers social affordances that are common in day-to-day life (we don't advertise these in person either).

The combination of Goffman's work and cognitive dissonance theory (Festinger, 1957) provides a useful framework for the design of behavior-change applications, but they do rely on some important assumptions—for example, that our self-control goals always increase our chances of goal achievement. Yet we all have experience with some level of rebelliousness. One of us (Dorian) believes that if a phone were to tell her to put down that brownie because the digital scale sent data to say that she was getting fat, she would probably break the phone and eat two brownies. Work on "ironic processes" (Wegner, Schneider, Carter, & White, 1987) has demonstrated that when trying hard *not* to think about something, people will think about it *more*.

Moreover, as mentioned in the chapter on positive emotions, denying negative feelings or enforcing positive thinking via affirmations can have damaging effects to wellbeing. Moreover, there will be cases in which some people's personal health goals are self-destructive (as in the case of those with anorexia, for example). As part of future research in personal informatics, behavior change, and positive computing, we will need to work on better understanding such complicating issues and their relationship to design and to find ways to devise designs that favor balanced and holistic approaches.

With regard to social influence as a motivator, many current-day apps and websites supply features that allow users to share their milestones and other personal data with others as a way of leveraging the motivational effects of social support and pressure. Of course, the impact of social pressure is not simply always good or always bad. Clearly, the effects of, say, trying to quit smoking may be different depending on whether you are doing it privately or publicly with your friends watching. Peer pressure is notorious in its connotations regarding teen behavior, such as drug use and risk taking. More generally, research by Sonja Lyubomirsky and others (Lyubomirsky & Ross, 1997; White, Langer, Yariv, & Welch, 2006) have highlighted the negative correlation between social comparison and happiness. Furthermore, studies by social psychologists have shown how negative influences can spread in a population (Christakis & Fowler, 2007).

At the same time, a multitude of successful mental health and wellbeing programs, including Alcoholics Anonymous, coming-out programs, and the SuperBetter resilience app, encourage the participant to connect with a sponsor or ally. Positive role models, mentors, and social proof can be just as productively influential as the negative variety can be detrimental. An ongoing research and practice question for positive-computing researchers will be: How can we design to promote positive social influence, while preventing the social validation and spread of destructive patterns?

The Delicate Issue of Persuasion

The psychological research on motivation described earlier makes it clear that motivation is a complex and multidimensional construct. Yet as technologists we often seek simplified models to facilitate practical design work. For example, B. J. Fogg's (2009) behavior model aims for simplicity in the name of practical application and suggests that humans are motivated by pain/pleasure, hope/fear, and social acceptance/social rejection and that designers can boost motivation by manipulating these. This model underpins work in persuasive technology, an area that Fogg describes as "the automation of behavior change."

The Fogg model is used for diagnosing interaction design problems, uncovering marketing opportunities, and encouraging small daily habits. Of course these are not the same as supporting psychological wellbeing, so we would be unwise to try and transplant Fogg's model blindly to the positive computing context. For example, highlighting motivators such as social acceptance and fear in ways that exploit low self-esteem or that contribute to constant anxiety (e.g., "Get rid of unsightly belly fat!," "We are under attack!"), although effective for many of the applications of persuasive technology, can have serious consequences for wellbeing. Furthermore, in positive computing we need to consider different *types* of motivation (e.g., extrinsic versus intrinsic) because research shows that motivation as it relates to wellbeing is not one size fits all.

Researchers are working to apply ethical guidelines to persuasive technology (as mentioned in chapter 4) and from the perspective of positive computing, we can view the issue as related to impact on wellbeing. There are various types of motivation and different ways of appealing to them, some of which improve wellbeing and some of which don't. In general, if wellbeing is our aim, as technologists we must be wary of basing our efforts on a simplified view of human beings and thus risking harm. This risk is one of the reasons we argue so strongly for multidimensional evaluation and for partnerships with wellbeing psychologists who will have a broader knowledge of human behavior and of the strengths and limitations of various models.

How exactly we define a range of motivational impact that includes helpful support at one end and manipulation and propaganda at the other will be a point of ongoing professional debate, but surely both transparency of motives and individual autonomy must be central to making the distinction.

As described in chapter 4, there are other approaches to supporting motivation for behavior change. One that has rapidly gained popularity among those working on population-wide wellbeing initiatives goes by the name of the book that popularized it: *Nudge*.

Nudging Positive Change—Designers as Choice Architects

Nudge theory can be applied to technology design, but it has been more famously positioned as a model for public policy with the distinct aim to improve organizations and society.

In the book that triggered the wave of interest to follow, Richard Thaler and Cass Sunstein (2008) describe their notion of "libertarian paternalism" that underscores the notion of nudging. A nudge, as they use the term, is "any aspect of the choice architecture that alters people's behavior in a predictable way without forbidding any options or significantly changing their economic incentives." In essence, it's about designing things such as policies and information in a way that favors healthier decisions, but without removing choice.

It is perhaps a shame that Thaler and Sunstein opted to use the term *paternalism* (which they openly acknowledge is saddled with negative and oversimplified connotations). For many, this term obscures the fact that Thaler and Sunstein insist on preserving choice and on basing assessments of improved life satisfaction on the *individual's* values rather than on the choice architect's values.

Perhaps the most compelling aspect of their argument is the notion that it is impossible to be neutral. When we design information technology (as is true for determining policy), we must base our decisions on *something*, and even if that something is no more than a coin toss, our decisions will have impacts on wellbeing. If that is the case, it seems crazy to base design decisions on mere chance when we might inform them with knowledge of what will improve people's lives. VSD might look at user values and determine ways of designing to improve user goals based on those values. Positive computing will look at psychological wellbeing and how we can make architectural decisions that are less likely to do psychological harm and more likely to do psychological good, without removing choice. What nudging reminds us is that for any work that seeks to improve lives, it is essential that liberty and autonomy are preserved and that notions of improvement are based on evidence, not on designer opinion (that evidence being research in psychology in the case of positive computing).

Engagement and Flow

In order to move from motivation to ongoing action, one must *engage* in some sort of activity, and, as such, these two factors are generally studied together. Engagement is quite influencable, so we can purposely design user interfaces and technological interventions that aim to engage people; in other words, our hopes for having impact in this area are well founded.

The Rules of Engagement

To have a prior engagement or be engaged means you have committed seriously to some person or purpose. Likewise, when we are engaged in a digital experience, we have committed to it, perhaps not consciously, but nevertheless it has seduced us to spend our valuable time with it. When educators discuss engagement, they're speaking about the holy grail of sustained student attention and interest. Positive psychologists often use the term *engagement* to refer to an active participation with one's community, with society, or with a higher purpose. Although this term can mean a variety of things, in this chapter we use it to refer to that combination of motivated commitment and sustained attention that educators, psychologists, media designers, and other architects of experience so often seek to elicit from their audiences.

Although there is no unilateral agreement on a precise definition or taxonomy of engagement, there tends to be consensus that when someone is engaged in an activity, they are more likely to enjoy it, to produce quality outcomes, and to learn more (Graham & Weiner, 2012). Most of the academic work on engagement has focused on student engagement in school and employee engagement on the job.

Researchers in the learning sciences (Christenson, Reschly, & Wylie, 2012) provide a definition of engagement that breaks it down into four types:

• **Emotional engagement** is assessed by detecting supportive emotions (e.g., interest) and the absence of negative, withdrawal emotions (e.g., anxiety or boredom).
• **Cognitive engagement** is demonstrated when the student uses sophisticated rather than superficial learning strategies.
• **Behavioral engagement** is generally assessed by observing concentration, attention, and effort (e.g., time on task). This type of engagement is the most straightforward to measure automatically—for example, by using

computer vision techniques or behavioral analytics based on the digital traces left during online activities.

• **Agentic engagement** is signaled by the student's active contribution to her learning experience (e.g., through creativity and innovation).

High levels of engagement are frequently described as being contingent on appropriate challenge, autonomy, and intrinsic motivation. Parvaneh Sharafi, Leif Hedman, and Henry Montgomery (2006) have mapped these attributes onto a multidimensional model they describe as the engagement modes (EM) model. The EM model specifies five engagement modes: (1) enjoying/acceptance, (2) ambition/curiosity, (3) avoidance/hesitation, (4) frustration/anxiety, and (5) efficiency/productivity. These modes are described on three dimensions: evaluation of object, locus of control between subject and object, and intrinsic or extrinsic focus of motivation. In the EM model, flow emerges when the user faces a pleasurable challenge that is also possible to master.

Flow—the Holy Grail of Engagement

Flow represents a state of total involvement in autotelic (intrinsically motivated) activity because the activity is so engaging it removes one from reflective self-consciousness. It's hard to imagine a game designer or software architect who wouldn't be thrilled to hear that their users were entering states of flow. This reason alone has made it worthy of study among technologists looking to design more rewarding digital experiences.

As part of the *Oxford Handbook of Positive Psychology*, Jeanne Nakamura and Mihaly Csikszentmihalyi (2009), the originator of flow theory, identify a number of requirements of flow experiences, which include perceived challenges that neither overwhelm nor underutilize our skills, clear reachable goals, and feedback that indicates if we are getting close to these goals. When these requirements are met, it is possible to get into "flow," an experience that combines:

1. *Concentration* on the present moment
2. Perceived *agency* over the situation or activity
3. A merging of action and awareness
4. A *loss of* reflective self-consciousness
5. A distortion of temporal experience
6. Intrinsic rewards

How many technological features are designed to increase productivity, but in ways that become a hindrance to flow experience? (Think of beeping phone notifications.)

Csikszentmihalyi (1997) describes flow as an "optimal experience" and a key to happiness. Thorough treatment of the varied affective and cognitive consequences of flow ranging from increased subjective wellbeing and greater life satisfaction to addiction on the extreme negative end are well beyond the scope of this book but are included in the collection *Advances in Flow Research* (Engeser, 2012).

Detecting and Measuring Engagement

Both emotional and cognitive engagement occur internally, and, as such, analysis has relied either on subjective reports or on instruments such as the Motivation and Engagement Scale (Martin, 2007), often used in pre/postexperimental designs, measurement studies, and path modeling. Both of these techniques have the limitations associated with after-the-fact reporting (Liem & Martin, 2012).

Flow has been measured through interviews, surveys, and experience sampling. More recently, advances in emotionally intelligent interfaces (Grafsgaard, Wiggins, Boyer, Wiebe, & Lester, 2013) and computer modeling allow us to approach the measurement of engagement constructs in new ways. In particular, sensing- and affective-computing techniques allow for ways of integrating observed expressions of engagement with internal self-reported measures. These techniques use sensors such as video cameras to record voices, facial expressions, and physiology (Calvo & D'Mello 2010). Moreover, we can generalize data from subjective reports in order to detect engagement automatically from a wider number of users more easily and with increasing accuracy.

Kazuo Yano, Sonja Lyubomirsky, and Joseph Chancellor (2012) describe an experiment in which flow was detected by physiological sensors as consistency of movement: "The key indicator of flow turned out to be consistency in movement. For some people, that consistent movement was slow; for others it was fast. Some were morning people; others favored the afternoon or evenings. Regardless, when participants experienced flow, their motions became more regular, as they lost themselves in a challenging but engrossing activity."

Engagement and Games

When we're not using our digital devices to engage with learning or work, there's a good chance we're using them to play. We've been playing games from before we could write (personally and historically), which suggests that games meet basic human psychological needs. One of the great powers of games is their ability to engage us fully. Gaming is an area in which levels of engagement can be particularly high, so digital game research is an obvious place to turn for guidance on engaging users in aid of wellbeing, but also for insight into when engagement becomes addiction and impacts wellbeing for the worse.

So do videogames increase or decrease wellbeing? According to research, the answer is: both. Our challenge is to pull apart the fibers of the digital gaming experience until we can better understand which characteristics improve wellbeing in what contexts and which don't.

When it comes to books, whether reading improves or decreases your wellbeing depends largely on content. With videogames, content is significant, but there are also critical wellbeing implications for how a game is played and who it's played with (Johnson, Jones, Scholes, & Colder Carras, 2013)—in other words, the social context and the game mechanics.

Exposure to violent videogames has consistently been shown to increase aggression, desensitize to violence, and reduce prosocial behavior. For example, a recent meta-analytic review (Anderson et al., 2010) concludes that "the evidence strongly suggests that exposure to violent video games is a causal risk factor for increased aggressive behavior, aggressive cognition, and aggressive affect and for decreased empathy and prosocial behavior."

But not all games are violent, and, unsurprisingly, just as practiced aggression can increase aggression, prosocial games can increase prosocial behavior (Gentile et al., 2009). It's not just the content that matters, but *how* you play. Game mechanisms vary both in general and in the details, and we suspect that key wellbeing differences lie in both core game mechanisms and in those easy-to-miss details.

The negative effects of videogames are fairly well publicized, but research on the positive effects of games on both psychological and physical wellbeing is less well-known. A survey of randomized controlled studies (Baranowski, Buday, Thompson, & Baranowski, 2008) on the use of videogame interventions for improving mental and physical health-related outcomes revealed moderate to strong positive results across 38 studies representing 195 health outcomes. These studies used videogames to provide physical therapy, psychological therapy, and improved disease self-management, among other things. Interestingly, positive effects were strongest (69 percent) for *psychological* outcomes.

Only seven of the studies were psychological interventions, and they included the use of casual, strategy, and custom games for post-traumatic stress disorder, anxiety, age-related cognitive decline, dyslexia, attention capacity, and self-esteem. Furthermore, many more studies in the literature were not included in this survey because they did not meet the randomized controlled trial criteria. The authors conclude that higher-quality studies remain relatively uncommon in this area and that "in order to most effectively assess the potential benefits of video games for health, it will be important for further research to utilize (1) [randomized controlled trial] methodology when appropriate; (2) longer follow-up duration; (3) improved measures of quality, such as randomization and blinding; and (4) standardized measurement tools and careful attention to the quality of outcome measures." In other words, early results are promising, but more work remains to be done.

Carmen Russoniello, Kevin O'Brien, and Jennifer Parks (2009) showed the ability of casual games to reduce stress, anxiety, and depression. As mentioned in chapter 6, a recent review of digital games (Johnson et al., 2013) looked at their impact on young people's wellbeing and concluded that they "positively influence young people's emotional state, self-esteem, optimism, vitality, resilience, engagement, relationships, sense of competence, self-acceptance and social connections and functioning."

This impact suggests that games are still largely untapped for positive-computing research and design. Of course, in the midst of our enthusiasm, we must still be careful to resist sweeping generalizations. That some studies have shown that some specific casual games can decrease stress and depression is very significant and means we can design for that outcome— but it doesn't follow that *all* games or even all casual games do this for all people or that secondary effects may not also emerge.

By way of anecdotal illustration, many of us have probably already enjoyed the stress-relieving qualities of *Bejewelled*, the popular game that proved therapeutic in the East Carolina study (Russoniello, O'Brien, & Parks, 2009). However, this same game has multiple incarnations, one of which, *Bejewelled Blitz*, is styled with all the bells-and-whistle rewards of slot machines and gambling. As enjoyable as I (Dorian) found it myself, I was persuaded to remove it from my tablet when my seven-year-old began begging me for money to use on gambling in the game. I shuddered to think it might be conditioning his formative mind to respond all the more rapidly to this kind of manipulation later in life.

The problem for wellbeing in this scenario is not *games* in general or even the core game mechanism itself (pattern matching), but the way in which this particular version was designed to profit financially from gambling-based incentives. When I replaced it with an equivalent that employs the same core pattern-matching mechanism, but without the gambling layer, *Candy Crush*, the negative side-effect disappeared.

My humble experience is hardly research evidence, but it is illustrative of how multilayered and complex the short- and long-term effects of even apparently simple casual games will be on wellbeing (as well as of how design for optimum profit can come up against design for wellbeing). Inevitably, *some* design aspects of any technology may contribute to wellbeing, while *others* may decrease it in parallel. This is what seems to occur with cooperative violent videogames that have been correlated with aggression but also cooperation skills (Greitemeyer, Traut-Mattausch, & Osswald, 2012; Velez, Mahood, Ewoldsen, & Moyer-Gusé, 2012). Playing as a team, users can learn valuable cooperation skills and find new ways to connect, but doing so in the context of enacting graphic violence or as part of a simplistic "us" versus "them" mentality are separate aspects with potentially negative consequences for the wellbeing of individuals and society. Game designer Raph Koster (2013) suggests that our gameplay (digital and otherwise) continues to reinforce instinctive skills once critical for our survival as cavemen but now obsolete, such as shooting and aiming (once important for hunting). He also cites "blind obedience to leaders and cultism, rigid hierarchies, binary thinking, the use of force to resolve problems, like seeking like and its converse xenophobia" as common themes based on obsolete skills. He suggests that we should be designing games that evolve with us and reinforce skills relevant to the modern world. By way of example, he highlights the game *Diplomacy*, which according to Koster is not only an example of a game that reinforces modern skills, but one that also provides "evidence that remarkably subtle interactions can be modeled within the confines of a rule set and traditional role-playing can reach the same heights as literature in the right hands."

Game addiction is another serious concern for anyone looking to use games to foster psychological wellbeing, and research has implicated risk factors such as personality traits, motivations for playing, and structural game characteristics (Kuss & Griffiths, 2011). But what if games themselves could be used to build resilience against game addiction while fostering positive engagement? Work uncovering risk factors (Kuss & Griffiths, 2011) and studies on resilience factors (risk of Internet addiction reduced by self-compassion [Iskender & Akin, 2011]) have begun to pave the way.

Although a discussion on addiction is well beyond this book's scope, the message emerges that if we are to be genuinely effective in leveraging the incredible potential of games for positive computing, we need to work carefully through what will be a slowly unraveling story of psychological impact and an important ongoing area of research. The key to a future of positive games lies in giving these technologies credit for producing highly multifaceted and complex experiences, acknowledging the incredible potential they provide, and exploring all the effects of game design on wellbeing so that we can increasingly favor the beneficial ones.

Design Implications

Designing to Motivate

The *Zombies, Run!* game, alluded to at the start of this chapter, is one of a multitude of apps and "exergames" whose primary offering is motivation—namely, motivation to do things we don't otherwise feel intrinsically motivated to do. We may want to do things that improve our wellbeing, but when those things also require effort or are unpleasant in the moment, we are at odds with the pleasure principle. Technology, sometimes via gamification, can step in to resolve the conflict. By layering experience and challenge that are intrinsically enjoyable (e.g., playing a part in a zombie narrative) over the activity that isn't (e.g., running), if the two activities are sufficiently intertwined (running becomes part of the story), then the whole experience can become more rewarding, thereby increasing our intrinsic motivation to take part.

Uplifted, created by the United Kingdom's Channel 4 for promoting positive emotions, takes a slightly different approach and embeds moments of positive reflection into an *Angry Birds*–style casual game. The game and reflection are thematically linked but not intrinsically linked as they occur separately and have little bearing on one another.

Other approaches engage our self-determined extrinsic motivation by helping us to articulate and track goals, be they larger goals (I will eat healthier) or smaller subgoals (I will chug a glass of water every morning) in aid of closing the gap between our behavior and our goals and values.

Designing for Engagement and Flow

In 2006, Yvonne Rogers (2006) discussed designing for increased engagement as an alternative to quiet automation for a future of ubiquitous computing. Rogers proposes "a significant shift from proactive computing to proactive people; where UbiComp technologies are designed not to do things for people but to engage them more actively in what they currently do. Rather than calm living[,] it promotes engaged living, where technology is designed to enable people to do what they want, need or never even considered before by acting in and upon the environment." Design for human autonomy (as opposed to machine autonomy) will be critical to fostering wellbeing (Calvo, Peters, Johnson, &

Rogers, 2014). An important balancing factor for human autonomy (as pointed out to us by Ben Shneiderman in conversation) is the reality of interdependence and our interconnected relationship to others. Both of these are core to SDT. If we were to look at design for motivation and engagement from the perspective of SDT, then when we evaluate a technology, we might ask these questions: Does the user experience respect autonomy? Does the user experience support a sense of competence? Does the user experience support connection to others? Of course, in being a theory of both wellbeing and motivation, it should also help us create conditions for engagement and flow.

Although it is generally accepted that there is no way to reliably design an experience of flow (the triggers are too individual), there are nonetheless ways to design conditions that increase the likelihood of flow experience. One approach to doing this is to identify the obstacles to flow and design to reduce them. Researchers in attentive computing investigate, among other things, how interruptions can be minimized and attention sustained. Modern versions of popular productivity software such as Microsoft Word and WordPress have incorporated options that remove screen clutter, allowing the user to focus only on the task at hand. Similar examples were Apple's inclusion of the "Do not disturb" setting in its iOS, and Freedom, a software application whose sole purpose is to allow you to easily shut down your Internet connection for a set period of time so you can proceed without distractions.

Another approach to supporting flow is to identify those conditions that are particularly conducive to it and include them where possible. For example, based on the work by Nakamura and Csikszentmihalyi (2009), we should design for appropriate challenge, clear reachable goals, and feedback that indicates if we are getting close to these goals. Certainly, these conditions are already familiar to game and interaction designers. A third approach to designing for flow, rather than looking at the universals, honors the individuality of what triggers flow for different people. In a study mentioned earlier (Yano et al., 2012), the research team used sensor data to help workers determine what time of day they were most likely to get into flow. They could then adjust their schedules and habits accordingly.

A significant characteristic of flow theory for HCI is its *interactionism* (a focus on a system made of the person and the environment). Csikszentmihalyi (1997) describes an emergent motivation that arises from the interaction that forms the context of a flow experience rather than a motivation that is a property of the individual (i.e., commitment) or of the environment (e.g., persuasion). Such a line of argument goes to the heart of current HCI views that combine interactionism and embodiment, such as those by Paul Dourish (2004).

The ways in which different users and technologies interact will be no less an issue for positive computing. For example, reminders to engage with one activity may increase distractions for others and reduce chances of engagement and flow conditions. There are surely ways around this, but the need to analyze the impact of design changes holistically is fundamental.

A final approach to supporting motivation and engagement lies in helping each user discover what unique conditions motivate him or her, which can be supported by behavioral analytics, personal feedback, and reflection.

In the next chapter, we turn to the self, looking at the various ways in which technology can support reflection and self-awareness. We also consider how these apparent virtues can flow unwittingly into negative experience such as rumination and narcissism and how we might be guided by a notion that is at once ancient and groundbreaking: self-compassion.

Note

1. The *Zombies, Run!* game consists of audio recordings and an accompanying website that together weave a zombie apocalypse narrative and elements of game mechanics around your exercise routine. Put simply, you'll run faster and enjoy yourself more if you're being chased by zombies. See zombiesrungame.com for more information.

References

Anderson, C. A., Shibuya, A., Ihori, N., Swing, E. L., Bushman, B. J., Sakamoto, A., ... Saleem, M. (2010). Violent video game effects on aggression, empathy, and prosocial behavior in eastern and western countries: A meta-analytic review. *Psychological Bulletin, 136*(2), 151–173.

Baranowski, T., Buday, R., Thompson, D. I., & Baranowski, J. (2008). Playing for real: Video games and stories for health-related behavior change. *American Journal of Preventive Medicine, 34*(1), 74–82.

Calvo, R. A., & D'Mello, S. (2010). Affect Detection: An Interdisciplinary Review of Models,Methods, and their Applications. *IEEE Transactions on Affective Computing, 1*(1), 18–37.

Calvo, R. A., Peters, D., Johnson, D., & Rogers, Y. (2014). Autonomy in technology design. In *CHI'14 extended abstracts on human factors in computing systeMs*. New York: ACM.

Christakis, N. A., & Fowler, J. H. (2007). The spread of obesity in a large social network over 32 years. *New England Journal of Medicine, 357*(4), 370–379.

Christenson, S. L., Reschly, A. L., & Wylie, C. (Eds.). (2012). *Handbook of research in student engagement*. New York: Springer.

Consolvo, S. (2009a). Designing for behavior change in everyday life. *IEEE Computer, 42*(6), 100–103.

Consolvo, S. (2009b). Theory-driven design strategies for technologies that support behavior change in everyday life. In *Proceedings of the 27th International Conference on Human Factors in Computing Systems* (pp. 405–414). New York: ACM.

Consolvo, S., & Everitt, K. (2006). Design requirements for technologies that encourage physical activity. In *CHI '06 proceedings of the SIGCHI Conference on Human Factors in Computing Systems* (pp. 457–466). New York: ACM.

Csikszentmihalyi, M. (1997). *Finding flow: The psychology of engagement with everyday life*. New York: Basic Books.

Dourish, P. (2004). *Where the action is: The foundations of embodied interaction*. Cambridge, MA: MIT Press.

Dweck, C. S. (2006). *Mindset: The new psychology of success*. New York: Ballantine Books.

Dweck, C. S., & Leggett, E. L. (1988). A social-cognitive approach to motivation and personality. *Psychological Review, 95*(2), 256–273.

Engeser, S. (Ed.). (2012). *Advances in flow research*. New York: Springer.

Festinger, L. (1957). *The theory of cognitive dissonance*. Stanford, CA: Stanford University Press.

Fogg, B. J. (2009). A behavior model for persuasive design. Paper presented at Persuasive 2009, 4th International Conference on Persuasive Technology. Claremont, CA. April 26–29.

Gartner says by 2015, more than 50 percent of organizations that manage innovation processes will gamify those processes. (2011). *Gartner Newsroom*, April 12. http://gartner.com/newsroom/id/1629214.

Gentile, D. A., Anderson, C. A., Yukawa, S., Ihori, N., Saleem, M., Ming, L. K., … Sakamoto, A. (2009). The effects of prosocial video games on prosocial behaviors: International evidence from correlational, longitudinal, and experimental studies. *Personality and Social Psychology Bulletin, 35*(6), 752–763.

Goffman, E. (1959). *The presentation of self in everyday life*. Garden City, NY: Doubleday.

Graham, S., & Weiner, B. (2012). Motivation: Past, present and future. In K.R. Harris., S. Graham., & T. Urdan (Eds), *APA educational psychology handbook* (pp. 367-398). APA: Washington.

Greene, D., & Lepper, M. R. (1974). Effects of extrinsic rewards on children's subsequent intrinsic interest. *Child Development, 45*, 1141–1145.

Grafsgaard, J. F., Wiggins, J. B., Boyer, K. E., Wiebe, E. N., & Lester, J. C. (2013). Automatically recognizing facial expression: predicting engagement and frustration. In S. D'Mello, S. K., Calvo, R. A., & Olney, A. (eds.), *Proceedings of the 6th International Conference on Educational Data Mining*. July 6–9, in Memphis, Tennessee, USA.

Greitemeyer, T., Traut-Mattausch, E., & Osswald, S. (2012). How to ameliorate negative effects of violent video games on cooperation: Play it cooperatively in a team. *Computers in Human Behavior, 28*(4), 1465–1470.

Iskender, M., & Akin, A. (2011). Self-compassion and Internet addiction. *TOJET, 10*(3), 215–221.

Johnson, D., Jones, C., Scholes, L., & Colder Carras, M. (2013). *Videogames and wellbeing: A comprehensive review* (pp. 1–37). Melbourne, AU: Young and Well Cooperative Research Centre.

Kayes, D. C. (2004). The 1996 Mount Everest climbing disaster: The breakdown of learning in teams. *Human Relations, 57*(10), 1263–1284.

Koster, R. (2013). *A theory of fun in game design*. Phoenix: Paraglyph Press.

Kuss, D. J., & Griffiths, M. D. (2011). Internet gaming addiction: A systematic review of empirical research. *International Journal of Mental Health and Addiction, 10*(2), 278–296.

Liem, G. A. D., & Martin, A. J. (2012). The Motivation and Engagement Scale: Theoretical framework, psychometric properties, and applied yields. *Australian Psychologist, 47*(1), 3–13.

Locke, E., & Latham, G. (2006). New directions in goal-setting theory. *Directions in Psychological Science, 15*(5), 265–268.

Lyubomirsky, S., & Ross, L. (1997). Hedonic consequences of social comparison: A contrast of happy and unhappy people. *Journal of Personality and Social Psychology, 73*(6), 1141–1157.

Martin, A. J. (2007). Examining a multidimensional model of student motivation and engagement using a construct validation approach. *British Journal of Educational Psychology, 77*(2), 413–440.

Maslow, A. H. (1943). A theory of human motivation. *Psychological Review, 50*(4), 370–396.

Miles, M. B. & Huberman, A. M. (1994). *Qualitative data analysis: An expanded sourcebook.* Thousand Oaks, CA: Sage.

Nakamura, J., & Csikszentmihalyi, M. (2009). Flow theory and research. In C. R. Snyder & S. J. López (Eds.), *Oxford handbook of positive psychology* (pp. 195–206). Oxford: Oxford University Press.

Reiss, S. (2004). Multifaceted nature of intrinsic motivation: The theory of 16 basic desires. *Review of General Psychology, 8*(3), 179–193.

Rogers, Y. (2006). Moving on from Weiser's vision of calm computing: Engaging ubicomp experiences. *Proceedings of the 8th International Conference of Ubiquitous Computing*, Orange County, CA, USA, September 17–21 (pp. 404–418). Heidelberg: Springer.

Russoniello, C. V, O'Brien, K., & Parks, J. M. (2009). The effectiveness of casual video games in improving mood and decreasing stress. *Journal of CyberTherapy and Rehabilitation, 2*(1), 53–66.

Ryan, R., & Deci, E. (2000). Intrinsic and extrinsic motivations: Classic definitions and new directions. *Contemporary Educational Psychology, 25*(1), 54–67. doi:10.1006/ceps.1999.1020.

Sharafi, P., Hedman, L., & Montgomery, H. (2006). Using information technology: Engagement modes, flow experience, and personality orientations. *Computers in Human Behavior, 22*, 899–916. doi:10.1016/j.chb.2004.03.022

Street, H. (2002). Exploring relationships between goal setting, goal pursuit, and depression: A review. *Australian Psychologist, 37*(2), 95–103. doi:10.1080/00050060210001706736.

Thaler, R. H., & Sunstein, C. R. (2008). *Nudge: Improving decisions about health, wealth, and happiness.* New Haven, CT: Yale University Press.

Velez, J. A., Mahood, C., Ewoldsen, D. R., & Moyer-Gusé, E. (2012). Ingroup versus outgroup conflict in the context of violent video game: The effect of cooperation on increased helping and decreased aggression. *Communication Research*, (August): 1–20.

Wegner, D. M., Schneider, D. J., Carter, S. R., & White, T. L. (1987). Paradoxical effects of thought suppression. *Journal of Personality, 53*(1), 5–13.

White, J. B., Langer, E. J., Yariv, L., & Welch, J. C. (2006). Frequent social comparisons and destructive emotions and behaviors: The dark side of social comparisons. *Journal of Adult Development, 13*(1), 36–44.

Yano, K., Lyubomirsky, S., & Chancellor, J. (2012). Sensing happiness. *IEEE Spectrum, 49*(12), 32–37.

8 Self-Awareness and Self-Compassion

It was late in his life, as he faced his executioners, that one of the world's greatest thinkers would conclude that death was preferable to giving up philosophy. To those in his midst he declared simply: "The unexamined life is not worth living." Socrates would go on to swallow hemlock for his crimes—crimes that would come to represent a pinnacle of human achievement.

A century earlier ancient India had yielded another of the world's greatest philosophers, Gautama Buddha, who said "You are what you think. All that you are arises from your thoughts. With your thoughts you make your world." The core Buddhist tenet summarized by the Four Noble Truths explains that in order to end our experience of suffering, we must first understand the workings of our mind. We must be able to recognize our thoughts, our emotions, our reactions, and their precursors, as they are the roots of our suffering.

Methods for self-examination, introspection, and self-awareness have since been refined by generations of philosophers, and more recently by modern Western psychologists from Sigmund Freud to Aaron Beck.

Now, in the second decade of the twenty-first century, cognitive behavioral therapy (CBT) has gone online: experience sampling solicits mood updates via smart phone, and the quantified-self movement inspires growing innovation in personal data analysis. But is it reasonable to expect any deeper understanding of our personal thoughts and emotions to arise from data? Can artificial intelligence, sophisticated machine-learning algorithms, and digital visualizations really help us to know ourselves better? If they did, would we be better for it?

In this chapter, we look at the state of the art in technology-mediated reflection, the varied approaches that have been applied to this endeavor, and the research emerging from its use. We also look at the foundations in psychology that underpin efforts toward greater self-awareness and how they are linked to increases in wellbeing. We look at both the potentials and the problems that may be associated with technological intervention into our inner lives, while deliberately evading the question, Would the Buddha have tweeted "feeling one with the universe" from under the Bodhi Tree?

Know Thyself

Why bother? Is self-knowledge merely philosophically virtuous or actually important to greater happiness and flourishing? Methods for CBT demonstrate that increased self-awareness can indeed improve wellbeing. In fact, a healthy chunk of the research and clinical work in modern psychology *relies* on promoting awareness of one's thoughts, emotions, and behaviors in order to effectively treat mental illnesses such as anxiety and depression.

Neuroscientist Richard Davidson (e.g., Davidson & Begley, 2012) has identified recognizable neurophysiological patterns associated with self-awareness. He has found that greater self-awareness is associated with increased activity in the insula (involved in consciousness, emotion, and body regulation) and that long-time meditators have larger insulas.

If self-awareness is important to personal growth and wellbeing, what can we do to develop it, and how can technology play a part in that development? The answer is most frequently sought in various forms of reflection, introspection, and mindfulness training. There is an impressive wealth of evidence for the effectiveness of mindfulness practice on wellbeing, including the work of Jon Kabat-Zinn and researchers at the Oxford Mindfulness Centre. Davidson cites neurophysiological evidence for the effectiveness of mindfulness practice both for increasing self-awareness and for tempering potential negative side effects that might accompany this awareness (e.g., hypersensitivity and anxiety). We reserve mindfulness, however, for the next chapter and deal with it as a factor of wellbeing in its own right. In this chapter, we focus on reflection.

According to twentieth-century philosopher, psychologist, and educational reformer John Dewey (2013), it is essential for education to develop the skills of reflection: "while we cannot learn or be taught to think, we do have to learn to think well, especially acquire the general habit of reflection." For Dewey, reflection is an act of reason that is often but not always directed to understanding the external world. He was also very willing to apply this objective rationality to more emotional concerns. "The meanings of *honesty, sympathy, hatred, fear,* must be grasped by having them presented in an individual's first hand experience." Applying Dewey's imperative to modern technology design suggests that systems intended to help people reflect on emotions should do so from the context of the user's own life experiences rather than from abstract concepts. Of course, emotions are just one category of things upon which one might introspect. There are at least three others.

Targets of Reflection

When we speak of self-awareness or reflection, there are various interpretations regarding what it is we are becoming aware of or upon which we are reflecting. Over time there have been proposals for various targets of reflective thought, which can roughly be synthesized into four main categories.

Cognitive awareness is what we believe we know about the world around us and our own lives. Understanding this aspect in ourselves is generally referred to as "metacognition," a term introduced by John Flavell in the late 1970s (Flavell, 1979). A great deal of research on how to build computer systems to support metacognition is available in the learning technologies literature.

Affective awareness involves awareness of our states of mind, in particular moods and emotions. Strategies and computer tools that help us track and reflect on our moods and emotions have been developed in the areas of psychotherapy and affective computing and as commercial tools. Some of these tools are discussed later in this chapter.

Experiential awareness is our awareness of the integrated aspects of cognition, affect, and behavior (including their external and internal triggers). The combination of these three aspects of "experience" (cognition, affect, and behavior) is at the core of CBT, which we describe in more detail later in this chapter.

Character traits, or those aspects of personality that are dispositional qualities, are a fourth set of mental aspects upon which we can reflect. We won't be focusing on this last category in this chapter because there is insufficient research to draw on at this stage. We delve more deeply into the cognitive, affective, and experiential in the next three sections.

Cognitive Awareness and Metacognition

Metacognitive skills, or the ability to "know what one knows," are important to many aspects of life experience, from setting realistic personal targets to self-regulating learning activity. A number of researchers have explored metacognition in the context of designing intelligent tutoring systems. These systems can support learning by supporting metacognition—for example, encouraging self-explanation (as a way of scaffolding reflection) and setting personalized goals. In addition to using metacognition strategies as a feature, other systems have focused on developing metacognitive skills per se, based on the principles for metacognitive tutoring proposed by John R. Anderson and his colleagues (Anderson, Corbett, Koedinger, & Pelletier, 1995).

Outside of the learning context, researchers have considered the impact of metacognitive skills on mental health (particularly in the context of cognitive therapies), linking them to wellbeing. We come back to these examples later in the chapter.

Affective Awareness and Emotional Intelligence

The literature on emotional intelligence (EI) has focused on affective awareness—namely, the skills required to recognize and regulate our emotions in a way that is consistent with a model of emotional functioning (Mayer & Salovey, 1995). We discussed emotional intelligence in chapter 3, but it's worth coming back to Daniel Goleman's (1998) five EI skills because they provide a useful breakdown of affective awareness as it might contribute to wellbeing. The five EI skills are:

- Self-awareness (the ability to recognize our own emotions)
- Self-regulation (the ability to control them)
- Motivation (passion for what we do)
- Empathy (the ability to recognize others' emotions, which is covered in depth in chapter 10)
- Social skills, in particular the ability to manage relationships with others

Most of the research on EI links it to concrete measures of success, such as productivity in the workplace. Nevertheless, there is also some initial evidence for a link between EI and psychological wellbeing. For example, a number of studies have looked at how well EI predicts wellbeing as defined by standard SWB measures. Emma Gallagher and Dianne Vella-Brodrick (2008) analyzed the impact of EI on SWB, removing the effect of other factors such as personality and sociodemographic variables (e.g., wealth). They analyzed measures of life satisfaction, positive and negative affect, social support, EI, personality, and social desirability from the self-reports of 267 adults. Their analyses showed that EI and social support as well as their interaction effects significantly predicted SWB. Further research on these links would be helpful in demonstrating whether interventions to develop EI also increase wellbeing.

Experiential Awareness and Reflection

Perhaps the most promising strategies for reflection are those that target the holistic relationships between thoughts, feelings, and behaviors. The development of such understanding is at the core of some mindfulness practices and cognitive therapy, both of which have been empirically linked to wellbeing.

It's worth pointing out the challenges of interdisciplinary research around a construct with as many varied interpretations as self-reflection. Each discipline looks at different aspects of reflection and takes a different focus. On the one hand, we might approach reflection, as Socrates, Dewey, and Peter Salovey have, as a *retrospective* dimension of thought. Even within this category approaches differ; Socrates's point of departure was moral philosophy; Dewey's interest was cognition and education; and other authors have studied reflection from the perspective of professional practice (e.g., Schön, 1983). In contrast, in mindfulness training and Buddhist psychology, reflection is generally rooted in the *present*. It is a present-moment self-awareness that observes thoughts, emotions, and behaviors as they arise, often in the context of deliberate contemplative practice.

The concept of reflection has been used in so many ways it risks becoming a bit of a catch-all. Therefore, in order for the concept to be useful to positive computing in a practical sense, it must first be contextualized. The adoption of any model of reflection for the development of technology will be shaped by (*a*) our motives for targeting reflection in the first place (What problem are we trying to solve? What activity are we trying to support?), (*b*) the development process we follow, and (*c*) the methods of evaluation we use.

By way of example, in the next section we discuss two contexts within which technologies can be (and have been) used to support reflection. The first is psychotherapy, in which reflecting on cognitive and behavioral patterns has been demonstrated to effect transformative experiences, especially for those with mental illness. In the second example, we discuss reflection at a more prosaic level and at a narrower granularity. We look at the practice of reflecting retrospectively on our daily behaviors, moods, and goals and at how new technologies, particularly in the area of personal informatics, can be used to collect behavioral data and scaffold the reflection process.

Reflection as a Strategy for Mental Health Treatment

Cognitive Behavioral Therapy
A number of psychotherapies focus on developing a self-understanding of the relationships between our thoughts, feelings, and behaviors. One of the most commonly used psychotherapies in the treatment of depression and anxiety is cognitive behavioral therapy. CBT was developed by Aaron T. Beck (Beck, Rush, Shaw, & Emery, 1987) and is closely related to rational emotive behavioral therapy developed by Albert Ellis (1973), both during the late 1950s. While Beck was treating patients using a psychoanalytic method, he began to notice how they, especially those with depression, often misinterpreted or had "cognitive distortions" relating to events in their lives. They would, for example, selectively obstruct certain thought processes or overgeneralize. For instance, a patient might interpret the fact that his spouse didn't kiss him good-bye in the morning as evidence that he is no longer loved.

CBT is essentially based on the concept that the way we think influences the way we feel, which in turn influences the way we behave. This process, thinking–feeling–behaving, is part of an internal communication that people can access through reflection. The client can discover the meanings of such processes, in particular the triggers of irrational thinking and follow-up feelings, with the therapist's help. The therapist often follows a Socratic questioning approach, scaffolding the client's reflection so that he evaluates his assumptions and can modify his thinking. This systematic questioning (a.k.a. "talk therapy") is combined with other activities such as role playing, writing in a diary, disputing irrational beliefs, and modifying language to be more positive, assertive, or playful.

Technology for CBT
Richard Layard, notable British economist and proponent of national wellbeing measures, has called on the UK government to invest both in increasing the provision of CBT and in teaching EI in schools in aid of improving national wellbeing. One strategy for increasing the reach of CBT programs in light of a shortage of trained therapists is through technology, and computers have already been successfully used to deliver such programs. In fact, computer-based CBT (CCBT) has been successful enough that in

the United Kingdom the National Institute for Clinical Excellence considers Internet-delivered CBT a viable way of treating patients, in particular those with anxiety.

In a Health Technology Assessment for the UK National Health System, Eva Kaltenthaler and her colleagues (2006) compared the clinical and cost effectiveness of a number of online CBT products to traditional approaches. According to their analysis of 20 randomized controlled trial studies, there is evidence that CCBT is as effective as therapist CBT for the treatment of phobias and panic and is more cost effective for depression and anxiety. Moreover, using CCBT in conjunction with a therapist can reduce therapist time required.

It's worth noting that we do not believe computers can reasonably replace human mental health professionals. Any such notion would reflect a lack of understanding regarding the complexity of mental illness and the mental health profession. We reject a people-replacement model of technology's role in mental health, not only because of technology's limited ability to be creative and empathic and to provide genuine human presence, but also because technology-based programs are largely generalized, incapable of making critical insights or adapting sufficiently to the nuance and variety in human personalities and circumstances. As such, technological systems are probably incapable of safely assisting with nontextbook, long-term, and life-threatening cases. Nevertheless, for certain types of mental health challenges, they could contribute significant help in, at least, the following four ways:

- As **complement to therapist treatment** for a richer, more consistent, and possibly shorter treatment phase
- As **follow-up and maintenance** after therapist sessions are complete
- As **triage** where a shortage of qualified mental health professionals face far more people in need of help than can possibly be seen in one-on-one, hour-long sessions. Technology might be used to support more mild cases, while immediately directing those who may be in danger to professional help.
- As **a wider net** for the many people who, although they are in need of professional help, do not seek it for many reasons such as stigma, fear, cost, and logistics. The anonymity and easy access provided by online programs can potentially foster flourishing in a much larger number of people. Some of these people may even proceed to seeking professional help once transitioned by such a process. Others may find the online program in itself successful in improving their lives.

To these ends, initial research is promising. In an Australian evaluation of a CCBT system (Mackinnon, Griffiths, & Christensen, 2008), three conditions were explored: the use of MoodGym, an Internet-based CBT intervention; the use of Bluepages, a website with information; and a control placebo group. Results showed that the Internet interventions reduced depression symptoms to a greater extent than the control group. This was true for a post-test, the 6-month follow-up, and a 12-month follow-up.

A UK randomized controlled trial of the efficacy of CCBT (Proudfoot, Ryden, Shapiro, Goldberg, & Gray, 2004) showed even stronger outcomes. The authors compared a commercial multimedia CCBT system (Beating the Blues[1]) with traditional treatment and found that "the computerised therapy improved depression, negative attributional style, work and social adjustment. … For anxiety and positive attributional style, treatment interacted with severity such that computerised therapy did better than usual treatment for more disturbed patients. Computerised therapy also led to greater satisfaction with treatment."

These computer-based systems are exceptions in that they have been evaluated in peer-reviewed studies. Hundreds of other CCBT apps, many available for a few dollars at the app store, do not have the benefit of such evaluations. Most tend to be augmented diaries, providing users with a way to record events. Thoughts pertaining to these events can sometimes be labeled—for example, as "unhelpful" or "sad." Other apps focus on specific activities (e.g., sleep, diet, and drinking).

For example, Drink Coach is an app developed by the Haringey Advisory Group on Alcohol, a UK group supporting those who suffer from alcohol misuse. The app focuses on scaffolding reflection on drinking habits. The user can record her alcohol consumption and the "risks" associated with it as well as set goals. The system tracks alcohol units and related calories consumed over time, and diary entries include fields for craving duration and intensity. The app also provides videos about mindfulness and breathing exercises that can help with cravings.

Panic Attacks is an app produced by myCBT Ltd. that focuses on anxiety disorders. It provides audio recordings designed to be calming, information on panic attacks, and a diary that helps challenge misinterpretations.

We hope to see more research and rigorous evaluation of these kinds of apps in the future so that we all can learn more about how best to design this kind of support.

Technology-Mediated Reflection for Wellbeing

The examples of CCBT in the previous section deal with mental health *treatment*. However, the focus of positive computing is mental health *promotion*. Of course, we have elected to spend significant time describing these e-therapy approaches in part because they are some of the most sophisticated and well-evaluated examples of technologies directed at psychological functioning that exist today, but also because you don't have to be ill to benefit from them. For example, it is not only the clinically depressed that find themselves having irrational thoughts, making overgeneralizations, or "catastrophizing." Most of us are prone to the kind of mental habits that in larger amounts and combined with other symptoms characterize clinical anxiety and depression. Even those of us who are mentally healthy can thus benefit from the exercises and practice of detecting cognitive distortions. Reducing these habits in the general population can be seen not only as a preventative measure that builds resilience to illness, but also as a promotional measure that improves the level of wellbeing in the population overall.

If we look at the taxonomy of Internet-based medical interventions (Barak, Klein, & Proudfoot, 2009), it's interesting to imagine how many of these interventions might be reformulated as promotional (rather than therapeutic) strategies. Moreover, how many of them might be incorporated into the very tools we already use in our everyday activities? Although the notion of promotional strategies incorporated into everyday software remains somewhat forward thinking, there are a handful of examples of dedicated promotional tools. Among them is Echo.

Echo is a smartphone application for recording everyday experiences and reflecting on them afterward, created in collaboration by researchers in California (Isaacs, Konrad, Walendowski, & Lennig, 2013). They conducted three system deployments with 44 users who generated more than 12,000 recordings and reflections, and they found that the activity supported by the system (which they call "technology-mediated reflection") successfully improved wellbeing. This study is instructive not only because it demonstrates an effective design for supporting reflection to promote wellbeing, but also because it serves as a model for rigorous evaluation of positive-computing technologies. The research team assessed results using four separate psychological metrics: the Subjective Happiness Scale, Satisfaction with Life Scale, Psychological General Well-Being Index, and the Mindfulness Attention Awareness Scale.

Echo is just one example of the ways in which many of the exercises that a therapist would use are translatable to online delivery. For example, role-playing (think videogames), writing reflections (as with Echo and writing tools that we analyze in the next section), disputing irrational beliefs (for example, by scaffolding the reasoning process online), and modifying language (Could we have a positive-computing spell checker for reflective practice?) are all therapeutic strategies amenable to technology support. Clearly, there is much room for innovation in this area, and we look forward to future examples of technologies for the treatment, prevention, and *promotion* of mental health.

Reflection versus Direct Instruction

Reflective approaches to self-improvement are particularly appropriate for technology intervention in that they avoid the path of giving direct prescriptive instruction, which is a risky approach for any generalized tool to take. Moreover, reflective feedback has greater potential in situations where a client's full story and context are not clear, which is almost always the case online.

In one of the sidebars in chapter 3, Harvard social media scholar danah boyd describes the digital street—a poignant reminder of how public our lives are and the difficult considerations that have arisen around this new reality. The lives we see in these digital streets are visible only in fragments. Even when we want to put these fragments to good use—for example, by identifying people at risk in order to point them in the right direction for help—our task is not trivial because we remain unaware of the full context of their situation. In these instances, reflective interventions, where a person is (*a*) encouraged to contextualize issues for themselves, (*b*) provided with information upon which to reflect, and (*c*) not given direct advice to take a specific action, may pose the best solution.

In a project with the Young and Well Cooperative Research Centre, we currently are exploring how a computer could automatically detect cognitive distortions in what young people write in blogs and on social networks. Using natural-language processing, we hope to detect expressions of *all-or-nothing thinking*, *overgeneralization*, *discounting the positive*, and *jumping to conclusions*. Technologies that can recognize cognitive distortions might form the foundation for tools that help people recognize these distortions for themselves. Needless to say, the careful design of these tools will be critical. No one, least of all teenagers, wants a virtual agent telling them what he should or shouldn't post on his wall. But creatively and respectfully applied, with deference to autonomy, values, participation, and preferences, such technologies just might promote greater wellbeing on unprecedented scale.

Reflection for Wellness and Wellbeing—Quantifying the Self

Personal informatics, personal analytics, quantified self, self-tracking—these are all terms that refer to technologies used by individuals to collect and analyze data about their behaviors (and sometimes about their moods or emotions).

The area has grown on multiple fronts in both business and academia. Independent software developers and entrepreneurs can take credit for the impressive speed at which it has advanced. Take Buster Benson—a software developer in California and a pioneering example of a user/developer combo who has immersed himself with gusto into the world of quantified self. Benson has been quantifying himself since 2000.[2] In one of his earliest forays, he tracked his state of mind using a "mood-o-meter," a system he used to log and publish information about his morale, his health, and his sleep in concert with data on his alcohol and caffeine consumption. Using this application, he would rate these variables on a scale of 1 to 10, describe the day's events in a short diary entry, and produce plots and visualizations for him and others to view.

He points out that others found the visualizations valuable because they could better judge when to approach him to ask for a favor, and he found himself paying attention to the way in which he was perceived by others. Over the years he has built and often commercialized many more personal informatics tools. His quest, he says, is "to find meaning," and he carries out this quest by exploring data.

On another leg of the quest, after beginning a personal diary in order to track how his moods changed over time, he created the website 750words.com. The website is similar to a blog but differs in its constraints and purpose. It has a much simpler interface, limits posts to 750 words, and is designed to produce reports from these personal journal entries. A simple report might read, "Rafael Calvo started at 7:10 pm and finished 470 words at 7:57 pm, for a total of 47 minutes of typing at 10 words per minute. Rafael Calvo was distracted 4 times while writing." Benson integrates the information from this tool with photo streams, geomapping, emotional state tracking, the number of unresponded emails he has sitting in his inbox, tweets received, and myriad other data streams to produce an unusually detailed public portrait of his personal life.[3]

In a recent seminar, Benson shared his conflicting opinions of self-tracking, sometimes viewing it simply as compulsive behavior, sometimes finding meaning in the data, and sometimes finding that although there is meaning in it, it's "hundreds of years away."[4] He also found that after years of trying to find numbers that better match to his internal reality, more generic labels (or Boolean scores) seem best suited to the job. In one of his apps, he uses factors such as sleep, physical activity, meaningful work, time spent with his son, and so on to produce a single average measure.

Although Benson's personal voyage isn't scientific research, it is instructive. The experiences of people such as Benson who have been "quantifying themselves" for such long periods of time can provide useful insights in the way that diary and case study methods have been successfully used in HCI research. Sure, his concern for privacy is clearly lower than average, or perhaps he's simply courageous in the name of computer science, but the result is an intriguing public experiment (performance artwork, even) that anyone can explore in order to reflect on the potential benefits, risks, limitations, inanity, or promise of the thoroughly quantified life.

Together with people like Benson, entrepreneurs and developers are putting together all kinds of apps that help people reflect on their behaviors in light of data collected about almost any aspect of their lives. At the website PersonalInformatics.org, you can find a catalog with hundreds of applications, from those for diet and exercise to those for tracking your sex life.

Some developers integrate GPS data into applications that calculate running or cycling itineraries, distance benchmarks, speed, and approximate calories burned. Some companies add a website or a custom gadget to the mix. The gadget-based business model pertains to some of the most commercially successful products, such as those offered by Fitbit (in March 2013 valued at more than $300 million).

Possibly the most significant personal tracking experiences are occurring online, where people view visualizations, interact with gamified motivational features, and share data and goals with others. In our lab, we are currently developing a set of tools that combine observational data (e.g., from health gadgets and traffic logs) with self-reported data (e.g., responses to a CES-D questionnaire or other psychological instruments), aiming for a more holistic view.

Reflection will continue to play a central role in our technology-supported efforts in supporting self-awareness. However, relentless in our determination not to neglect the caveats, we know this chapter would not be complete without the case *against* too much reflection as a method for self-awareness to improve wellbeing.

Staring at Our Own Reflection

Can self-awareness (or at least reflection) go too far? When does healthy reflection become unhealthy rumination or obsessive self-focus? Will encouraging reflection simply feed the apparent rising trend in narcissism that is also linked to depression? Should we encourage self-focus when self-focus is known to be a conspicuous characteristic of depression? In a linguistic analysis of student essays, Stephanie Rude, Eva-Maria Gortner, and James Pennebaker (2004) found that depressed individuals used the word *I* more frequently than did their nondepressed counterparts.

These issues highlight the challenges inherent in supporting the "right kind" and "right amount" of reflection. As a way forward, we summarize some of the elucidating research on self-compassion and gratitude, potential antidotes to the risks of overreflection.

Rumination and Self-Criticism
Sonja Lyubomirsky, among many other psychologists, has studied rumination and its impact on depression, problem-solving ability, and sociability. While comparing people who tend to be happier with those who tend to be unhappier to the average, positive-psychology researchers have found a link between unhappiness and too much self-reflection, including "dwelling" (or rumination) and self-criticism. According to Lyubomirsky's (2001) research, happier people tend to self-reflect about moods and outcomes less than unhappier people.

Rumination, in which thinking is mostly about past events (rather than the worrisome future), tends to focus on issues of loss and bereavement, self-worth, and so on. Myriad studies have consistently shown that rumination is associated with depression and other mental health problems and that focusing attention on oneself can also increase and extend the length of depressive episodes. In the article "Rethinking Rumination" (Nolen-Hoeksema, Wisco, & Lyubomirsky, 2008), the authors explain how rumination is not just correlated with neuroticism, perfectionism, and other negative cognitive styles but also mediates "the relationship between depression, neuroticism, negative inferential styles, dysfunctional attitudes, self-criticism, dependency and neediness."

However, according to this same article, it is also not clear which forms of self-reflection can be adaptive, positive, and instrumental in personal change or even just benign. Some researchers have tried to separate factors using the Ruminative Responses Scale and other scales in attempts to disentangle positive forms of reflection from otherwise negative rumination. Evidence suggests a difference between analytic and experiential self-focus (Watkins & Teasdale, 2004), the latter relating to the nonjudgmental present-focused observation found in mindfulness practice or what Susan Nolen-Hoeksema and her colleagues (2008) call "concrete rumination" or "mindful experiencing." (We discuss mindfulness in the next chapter.)

This is perhaps why the generally present-focused and experiential process of self-reporting moods seems to increase emotional self-awareness, an important positive aspect of a number of psychotherapies. The positive impact of self-reports has been shown in studies where participants reported their moods via phone messages (Kaur et al., 2012) or as longer personal diary entries such as blogs (Ko & Kuo, 2009). Interestingly, the evidence suggests that mobile phone self-reports can help *reduce* the brooding component of rumination (which is the component with strongest correlation to depression) (Kaur et al., 2012).

Future research in positive computing can look at how technologies might be designed in such a way to detect signs of overthinking, brooding, or self-criticism and perhaps to shift focus to supporting antidotal practices such as mindfulness, connection to others, and change of perspective.

Self-Awareness versus Narcissism

Sometimes the barrage of inane status updates, blathering blogs, and twittering tweets conspire to create a cognitive and emotional cacophony that have driven many to one conclusion: the connected world is raising a Generation Me that believes everything they think and do is important and worthy of public display. These new tools begin to look less like they're about connecting and more like they're about performing.

To be fair, experimentation with any new tool will involve doing so clumsily at first. And certainly, we have already begun to adjust the way we use social media in light of these effects—for example, by filtering more, storing less, demanding finer privacy settings, and reporting spam. Western capitalist culture, especially in the United States, is frequently criticized for an overfocus on individualism, self-interest, and an overemphasis on self-esteem. A recent *New York Times* article reported that "Rutgers researchers classify 80 percent of Twitter users as 'meformers' who tweet mainly about themselves" (Tierney, 2013). Will tools for self-awareness and self-tracking simply reinforce this pattern, leading to a new wave of digital *selfing*?

Indeed, there's compelling evidence that both narcissism and narcissistic personality disorder have increased among younger generations in the United States (Twenge & Foster, 2010; Twenge, Konrath, Foster, Campbell, & Bushman, 2008). This downside comes, perhaps predictably, on the tail of more positive increases of other individualistic traits such as self-esteem, agency, assertiveness, and extraversion. Of course, you could say that previous generations had personal experience with economic depression, world war, and civil rights battles, which probably did more to temper a sense of entitlement than microblogging and personal digital devices do. But the meta-analyses look at levels of narcissism among college students from the early 1980s to the present day, not among their grandparents. Although the causes are surely various (from changes in education to cultural attitudes, politics, and lifestyles—all potentially playing a part), if we speculate that our digital technologies (or perhaps, more accurately, the ways we have designed and used them thus far) are playing some role in increasing levels of narcissism (and even if they're not to blame), it's sensible to ask what we can do about it in the context of positive computing. Contemporary researchers such as Paul Gilbert and Kristin Neff are among those who have spent the past decade investigating the potential of self-compassion to provide a more balanced way forward. While self-compassion might at first blush sound like just another "selfism," it critically incorporates a broader perspective and a sense of shared humanity that, combined with gratitude, may be a key strategy for keeping reflection in balance.

Self-Esteem versus Self-Compassion

If you grew up with Barney the purple dinosaur, you'll remember the theme song: "Cause you are special, special, everyone is special, everyone in his or her own way. ..." Dorian recalls her father scoffing cynically at the oxymoronic notion that everyone is special. Of course, growing up in a working-class family on the rural outskirts of Pittsburgh meant that if you wanted to feel special, you had to work for it. We now know young children benefit greatly from praise, encouragement, and affirmation of their competence and potential, which is why research on self-esteem has been incredibly important to the way educational institutions now prepare children for success in life. Still, according to some, there has been a downside to the approaches we've taken to boosting self-esteem. Described charmingly as the "Lake Woebegone Effect" (in reference to radio personality Garrison Keillor's fictional hometown, where "all women are strong, all men are good-looking, and all children are above average"), researchers have found that self-esteem that is contingent on external achievement or dependent on proof of perpetual above-averageness can lead to depression or narcissism down the track.

Psychologist Kristin Neff (2011) at the University of Texas, Austin, draws attention to the American obsession with needing to be above average. "In our incredibly competitive society, being average is unacceptable. We have to be special and above average to feel we have any worth at all. The problem, of course, is that it is impossible for everyone to be above average at the same time." She cites research that shows how we struggle to maintain the fragile sense of specialness required for our self-esteem by inflating our self-evaluations and putting others down to feel superior. It's hard not to think of all the "reality" TV shows and gossip magazines that in their condescending parade of the sensational sell us the opportunity to feel superior to others.

Neff goes on to unravel the consequences of a societal love affair with self-esteem, referencing grade inflation and the problems with a construct that is frequently contingent on external measures such as appearance, academic achievement, work performance, and social approval. "Contingent self-esteem drives people to obsess about the implications of negative events for self-worth, making them more vulnerable to depression and reduced self-concept clarity." Although she emphasizes that there are certainly healthy forms of self-esteem and there is much research linking self-esteem to wellbeing, she argues that these benefits can be found without the downsides in a notion of self-compassion.

Paul Gilbert (2009, 2010) introduced compassion-focused therapy in the past decade as a novel way to help people who suffer from high levels of shame and self-criticism. The concept of self-compassion may be new to Western psychology, but it is certainly not new to humanity. Gilbert credits Buddhist psychology as a source for compassion-focused therapy because it centers on compassion in its practice (Buddhist loving-kindness meditation begins with compassion for oneself).

According to Neff (2011), "self-compassion entails three main components which overlap and mutually interact: Self-kindness versus self-judgment, feelings of common humanity versus isolation, and mindfulness versus over-identification." Self-compassion has been correlated to increased wellbeing in multiple ways. It has also been shown to be a highly effective predictor of quality of life (Van Dam, Sheppard, Forsyth, & Earleywine, 2011) and has been correlated to other aspects of interest to positive-computing work, such as increased self-improvement motivation and reduced risk of Internet addiction. Laura Bernard and John Curry (2011) provide a summary of wellbeing correlates and suggest intervention strategies that could begin to inspire work in positive computing.

Gratitude and Appreciation
For a final antidote to over-focus on the self, and one that also fosters wellbeing in its own right, we turn to gratitude. The practice of gratitude in various forms (from thank-you letters to gratitude journals) has consistently been shown to increase wellbeing (for reviews see: Emmons & McCullough, 2004; Watkins, Woodward, Stone, & Kolts, 2003; Wood, Froh, & Geraghty, 2010). Christine Carter (2011), Director of Berkeley's Greater Good Science Center, recommends gratitude practice for curbing a sense of entitlement in our children as well as for fostering positive relationships: "Our culture glorifies independence and undervalues how much others help; we see our blessings as hard earned ... appreciation is one of the most important ways that we teach our kids to form strong relationships with others ... expressing gratitude is about expressing just how deep those connections run."

Design for gratitude already makes a few appearances in the virtual world. There are apps available to support gratitude practices (such as gratitude journals), games occasionally include opportunities for gratitude (e.g., *Hay Day* allows players to send thank you cards when help is provided by other players in the game), and the Learning Solutions website transformed the ubiquitous "like" into the more gratitude-focused "I appreciate this." Creative thinking around how we can support users in experiencing and extending acts of gratitude and appreciation will be a rewarding area of ongoing exploratory practice and one that can contribute positively to wellbeing.

A Way Forward
Tracking and sharing personal data can encourage us to compare ourselves to others, yet the research on happiness and self-compassion show that this can negatively impact our wellbeing. Reflecting on our thoughts, emotions, and behaviors is essential to personal growth, but obsessive rumination contributes to depression. How do we design technologies that support practices beneficial to wellbeing without reinforcing associated problems? We do not claim to have the answer, but we suggest the following design principles as safeguards in favor of healthy balance.

Design Implications

Principles for Supporting Self-Awareness and Reflection

- **Understand the pitfalls.** Cultivating an awareness and understanding of the pitfalls that exist surrounding reflection and self-tracking will help us (designers) to avoid inadvertently supporting them.
- **Design for self-compassion, gratitude, and mindfulness.** Research suggests that underscoring our efforts with principles of self-compassion, gratitude, and mindfulness will help prevent comparison, self-criticism, narcissism, and entitlement.
- **Acknowledge the limitations of technology and lean toward reflective support.** Honoring the diversity and complexity of people while acknowledging the limitations of what can be provided by technology will often mean employing reflective feedback and avoiding highly prescriptive and constrained solutions. Outside of human-mediated medical intervention, what we don't know about our users and their contexts will always outstrip what we do know, so providing them feedback for making better decisions will often be more widely effective than providing specific instruction.
- **Allow for nonabsolute categories.** When we present the analysis of personal data to support inferences about states of mind or behavior, we should not be constrained to predefined categories. This idea is based on the work by Ellen Langer, who showed the negative impact of preconceptions on creativity. In her studies, Langer (e.g., Langer & Piper, 1987) provided evidence that when people come to a task with strong, absolute conceptions, for example when they are told, "This is a ___," in contrast to a conditional conception, as when they are told, "This could be a ___," they are much less likely to adapt or be creative with the concept.

Follow the Ongoing Research

Although better understanding of how we can design safeguards and adaptations to ensure that wellbeing interventions genuinely promote wellbeing will require ongoing research, understanding that these tensions exist and following research in antidotal concepts such as self-compassion, gratitude, altruism, and sympathetic joy are sure to be critical to approaching an optimal balance over time.

And now, after alluding a number of times to the incredible promise of mindfulness for wellbeing, we turn to this unique state and practice in the next chapter—to its definition, its many positive correlates, and the strategies that have been proven to promote it.

Expert Perspectives—Technology for Self-Awareness

> **How Emotional Intelligence Can Inform Positive Computing**
>
>
>
> David R. Caruso, Yale University and EI Skills Group
>
> How you feel influences how you think and what you think about. Our decisions can be informed by our emotions and feelings, and they can also be derailed by them. Knowing how we feel, employing these feelings to assist our thinking, understanding the reasons we have these feelings, and successfully managing our feelings are the four key components of a theory proposed by Peter Salovey and Jack Mayer in 1990 called the "ability model of emotional intelligence (EI)."
>
> Although there are many popular conceptions of EI (sometimes referred to as Emotional Quotient or "EQ"), in the ability model EI is a form of intelligence related to analytical intelligence (IQ), but also different enough to be considered unique. Just like IQ, EI can be measured in an objective manner, and it shows individual differences: some people are more emotionally intelligent than others.
>
> This approach to EI posits that emotions are data, communicate meaning, and have various, predictable trajectories. Although we are often told that emotions can highjack us and derail us, emotions are adaptive. They help us survive, grow, and develop. A rich experience of emotions is essential to quality relationships. Although all emotions can be adaptive—including anger, fear, and disgust—EI defined and measured as an ability can inform positive computing.

(continued)

> The first ability in this model is perceiving emotions accurately. Positive-computing applications that help people better perceive their own emotions and the emotions of others would greatly benefit people who tend to misperceive emotions. Misunderstandings can be avoided through emotion recognition.
>
> The second ability is complex and includes the ability to generate emotions to help you feel what others feel, to connect with them, and to develop emotional empathy for others. It also allows us to use our emotions deliberatively in order to enter into a mood state that alters the way we think. For example, a sad mood may facilitate the detection of errors. A happy mood may inspire us to generate a large number of new ideas.
>
> Understanding the causes of our emotions and predicting how they might change based on various actions represents the third ability. Although emotions can often seem chaotic, they have rules. We are happy when we gain something of value, and we are sad when we lose something of value. What we value, of course, varies from person to person, but the underlying rules of emotions are fairly universal.
>
> Finally, if we perceive emotions of others, feel what they feel, understand how they might react to an event, what do we do with these emotions? Without the ability to manage emotions, we might be easily overwhelmed. The last and perhaps most important ability is managing our own and others' emotions.
>
> Identifying that we are on a trajectory toward sadness or disappointment and understanding that such emotions will not facilitate the task at hand, we can be given suggestions on how to alter the emotions, prevent unwanted emotions, maintain our emotions, or enhance our emotions. In essence, positive-computing systems can help us develop our EI capabilities.

Notes

1. Beating the Blues can be found at ultrasis.com.

2. Buster Benson discusses why he self-tracks in the video at goo.gl/Ds1kY.

3. See Benson's website at busterbenson.com.

4. See http://quantifiedself.com/2012/12/buster-benson-why-i-track/.

References

Anderson, J. R., Corbett, A. T., Koedinger, K. R., & Pelletier, R. (1995). Cognitive tutors: Lessons learned. *Journal of the Learning Sciences*, 4(2), 167–207.

Barak, A., Klein, B., & Proudfoot, J. G. (2009). Defining Internet-supported therapeutic interventions. *Annals of Behavioral Medicine*, 38(1), 4–17.

Beck, A. T., Rush, A. J., Shaw, B. F., & Emery, G. (1987). *Cognitive therapy of depression*. New York: Guilford Press.

Carter, C. (2011). *Raising happiness: 10 simple steps for more joyful kids and happier parents*. New York: Ballantine Books.

Davidson, R. J., & Begley, S. (2012). *The emotional life of your brain: How its unique patterns affect the way you think, feel, and live—and how you can change them*. New York: Hudson Street Press.

Dewey, J. [1910] (2013). *How we think* (p. 301). London: Forgotten Books.

Ellis, A. (1973). *Humanistic psychotherapy: The rational-emotive approach*. New York: The Crown Publishing Group.

Emmons, R. A., & McCullough, M. E. (Eds.). (2004). *The psychology of gratitude*. New York: Oxford University Press.

Flavell, J. H. (1979). Metacognition and cognitive monitoring: A new area of cognitive-developmental inquiry. *American Psychologist*, 34(10), 906.

Gallagher, E. N., & Vella-Brodrick, D. (2008). Social support and emotional intelligence as predictors of subjective well-being. *Personality and Individual Differences*, 44(7), 1551–1561.

Gilbert, P. (2009). Introducing compassion-focused therapy. *Advances in Psychiatric Treatment*, 15(3), 199–208.

Gilbert, P. (2010). An introduction to compassion focused therapy in cognitive behavior therapy. *International Journal of Cognitive Therapy*, 3(2), 97–112.

Goleman, D. (1998). *Working with emotional intelligence*. New York: Bantam.

Isaacs, E., Konrad, A., Walendowski, A., & Lennig, T. (2013). Echoes from the past: How technology mediated reflection improves well-being. In *CHI 2013: Proceedings of the SIGCHI Conference on Human Factors in Computing Systems* (pp. 1071–1080). New York: ACM.

Kaltenthaler, E., Brazier, J., De Nigris, E., Tumur, I., Ferriter, M., Beverley, C., ... Sutcliffe, P. (2006). Computerised cognitive behaviour therapy for depression and anxiety update: A systematic review and economic evaluation. *Health Technology Assessment*, *10*(33), 1–168.

Kaur, S. D., Reid, S. C., Crooke, A. H. D., Khor, A., Hearps, S. J. C., Jorm, A. F., ... Patton, G. (2012). Self-monitoring using mobile phones in the early stages of adolescent depression: Randomized controlled trial. *Journal of Medical Internet Research*, *14*(3), e67.

Ko, H. C., & Kuo, F. Y. (2009). Can blogging enhance subjective well-being through self-disclosure? *Cyberpsychology & Behavior*, *12*(1), 75–79.

Langer, E. J., & Piper, A. I. (1987). The prevention of mindlessness. *Journal of Personality and Social Psychology*, *53*(2), 280–287.

Lyubomirsky, S. (2001). Why are some people happier than others? The role of cognitive and motivational processes in well-being. *American Psychologist*, *56*(3), 239–249.

Mackinnon, A., Griffiths, K. M., & Christensen, H. (2008). Comparative randomised trial of online cognitive-behavioural therapy and an information website for depression: 12-month outcomes. *British Journal of Psychiatry 192*(2), 130–134.

Mayer, J. D., & Salovey, P. (1995). Emotional intelligence and the construction and regulation of feelings. *Applied & Preventive Psychology*, *4*(3), 197–208.

Neff, K. D. (2011). Self-compassion, self-esteem, and well-being. *Social and Personality Psychology Compass*, *5*(1), 1–12.

Nolen-Hoeksema, S., Wisco, B. E., & Lyubomirsky, S. (2008). Rethinking rumination. *Perspectives on Psychological Science*, *3*(5), 400–424.

Proudfoot, J., Ryden, C., Shapiro, D. A., David Goldberg, I. M., & Gray, J. A. (2004). Clinical efficacy of computerised cognitive-behavioural therapy for anxiety and depression in primary care: Randomised controlled trial. *British Journal of Psychiatry*, *185*(1), 46–54.

Rude, S., Gortner, E., & Pennebaker, J. (2004). Language use of depressed and depression-vulnerable college students. *Cognition & Emotion I*, *18*(8), 1121–1133.

Schön, D. (1983). *The reflective practitioner*. London: Temple Smith.

Tierney, J. (2013). Good news spreads faster on social media. *New York Times*, March 18.

Twenge, J. M., & Foster, J. D. (2010). Birth cohort increases in narcissistic personality traits among American college students, 1982–2009. *Social Psychological and Personality Science, 1*(1), 99–106.

Twenge, J. M., Konrath, S., Foster, J. D., Campbell, W. K., & Bushman, B. J. (2008). Egos inflating over time: A cross-temporal meta-analysis of the Narcissistic Personality Inventory. *Journal of Personality, 76*(4), 875–902, discussion 903–928.

Van Dam, N. T., Sheppard, S. C., Forsyth, J. P., & Earleywine, M. (2011). Self-compassion is a better predictor than mindfulness of symptom severity and quality of life in mixed anxiety and depression. *Journal of Anxiety Disorders, 25*(1), 123–130.

Watkins, E., & Teasdale, J. T. (2004). Adaptive and maladaptive self-focus in depression. *Journal of Affective Disorders, 82*(1), 1–8.

Watkins, P. C., Woodward, K., Stone, T., & Kolts, R. (2003). Gratitude and happiness: Development of a measure of gratitude and relationships with subjective well-being. *Social Behavior and Personality, 31*(5), 431–451.

Wood, A., Froh, J., & Geraghty, A. (2010). Gratitude and well-being: A review and theoretical integration. *Clinical Psychology Review, 30*(7), 890–905.

9 Mindfulness

The faculty of voluntarily bringing back a wandering attention, over and over again, is the very root of judgment, character, and will. ... An education which should improve this faculty would be the education par excellence.
—William James, *Psychology* (1892)

You get dressed, spill some coffee, put the cereal away in the fridge, pretend you're listening to your kids, and leave without the car keys. It's a typical day in the world of the modern *Homo sapiens*—a species that has largely lost its natural state of present awareness. We live on autopilot, lost in plans and reruns. Our attention flies off like a coven of witches to all the things we have to do, reliving the conversations we've had or rehearsing the ones we haven't. How many of us are mentally "at work" when we're eating dinner, lying in bed, or even on vacation? Worst of all, our digital devices seem to conspire against us. Picture this: it's the end of a long day, and you're finally unwinding—just beginning to pay real attention to what your spouse was saying as you settle back into the reality of the present. Suddenly, the vibrating beep attacks, you fail to resist, and it's an email message that flings you back to work, with a dose of stress for the ride.

The antidote or opposite state to all this "mind wandering" and "absent-mindedness" is *mindfulness*. Mindfulness, a kind of broad and nonjudgmental present-moment awareness, is not only a stress reducer, but also a key to wiser decisions, greater life satisfaction, and overall psychological and physical wellbeing, according to research. In fact, it's a bit of a magic bean; plant it, and a wealth of mental resources will be yours. That's probably why mindfulness practice is common to many religious traditions, especially Buddhism.

The Buddha was arguably the first person to prescribe a mental health intervention. It was a universal eight-step program he called the

Eightfold Path designed to support ultimate wellbeing, and among the steps is mindfulness.[1] Buddhist traditions have spent the past 2,500 years cultivating mindfulness through specific training practices (i.e., Vipassana, Zazen, and walking meditation), which is why researchers such as Mark Williams and Jon Kabat-Zinn (2013) have turned to these practices to see how they might be tested empirically and integrated into mindfulness interventions for health care, mental illness, and education.

Mindfulness practices were first adapted for a Western audience around the 1970s. Most significantly, in 1979 Jon Kabat-Zinn of the University of Massachusetts Medical School introduced the Mindfulness-Based Stress Reduction (MBSR) program. MBSR is an eight-week course that has since helped thousands of people with stress, pain, and depression (Kabat-Zinn, 1990) and that research has consistently shown to have large effect sizes (Grossman, Niemann, Schmidt, & Walach, 2004).

Based on MBSR, researchers Zindel Segal, John Teasdale, and J. Mark Williams, director of the Mindfulness Centre at the University of Oxford (see his sidebar in this chapter) developed the Mindfulness-Based Cognitive Therapy (MBCT) program, an integrated approach (now endorsed by the UK National Institute of Clinical Excellence) that has proven effective in treating depression, preventing relapse, and promoting wellbeing (see Segal, Williams, & Teasdale, 2012).

Privy to the findings on its many benefits, human-resources departments are now lining up to include mindfulness training as part of professional development. This increased attention has meant increased examination in the research literature as well, which is reflected in an overwhelming number of peer-reviewed studies each year. (The mindfulness research guide[2] has cataloged more than 2,500 papers, which began as a handful in 2000 and grew to almost 500 a year by 2012.)

The scientific evidence for the positive impact of mindfulness on wellbeing has been accumulating within neuroscience as well as within psychology and psychiatry. In this chapter, we look at some of the literature in each of these fields and present examples of the kinds of traditional and technology-based strategies already in place for developing mindfulness. We also look at some of the latest technologies emerging to support mindfulness practice. But, first, the requisite foray into the workings of the mind or how, from our thoughts to our brainwaves, we change measurably as a result of sustained mindful attention.

The Psychology of Mindfulness

Awareness and Attention

Prominent Zen teacher and author Thich Nhat Hanh (2008) describes mindfulness as "keeping one's consciousness alive to the present reality." It is in fact two elements of consciousness—attention and awareness—that are at the core of mindfulness practice. Father of Western psychology William James (1892) conceived of mindfulness as the state of being attentive and aware of what is happening at the present moment. Likewise, Kirk Brown and Richard Ryan (2003) discuss mindfulness as a naturally occurring characteristic, with attention and awareness as its two components.

Awareness is the component that continuously monitors the inner and outer world. It is what feeds that which Daniel Kahneman (2013) has called "system 1," the automatic system. Attention refers to the focused attention we place on part of the moment-to-moment experience, which may be our thoughts, feelings, behaviors, or the surrounding environment. So depending on the moment itself, a focus of our awareness might be a flower, a companion, or a current feeling of anxiety. Attention provides a heightened sensitivity to a subset of what we sense. An essential element of this attention in the context of mindfulness is nonjudgment.

Mindfulness as Nonjudgmental Attention

Mindfulness is distinct from the retrospective reflective processes discussed in the previous chapter in at least two ways. First, rather than being a mental account or an *analysis* of experience, mindfulness is a nonreflective and *nonjudgmental* observation of it. Mindfulness practice specifically avoids evaluation, opinion construction, or analytical cognitions. It is attention and awareness stripped bare of the judgments we so often automatically impose upon all that we perceive. Kabat-Zinn (2003) makes this clear and offers the following as an operational definition of mindfulness accordingly: "the awareness that emerges through paying attention on purpose, in the present moment, and nonjudgmentally to the unfolding of experience moment by moment."

The nonjudgmental, nonreactive aspect of mindfulness training is critical in clinical settings, where it prevents attention from turning to thoughts of comparison and criticism (e.g., "I'm no good at this; I get distracted all the time" or "He's so much better than me"). The nonjudgmental aspect is also sometimes referred to as "detached" awareness (Speca, Carlson, Goodey, & Angen, 2000). Pema Chödrön (2007) describes it in terms of unconditional friendliness: "Unconditional friendliness is training in being able to settle down with ourselves, just as we are, without labeling our experience as 'good' or 'bad.'"

This relinquishing of judgment is also a key difference between mindfulness and CBT. CBT encourages the labeling of negative thoughts and feelings, for example, as "unhelpful" or "irrational" and encourages challenging and changing those thoughts. In contrast, mindfulness practice encourages the simple observation and acceptance of them and discourages both rumination and striving for any particular outcome (Hamilton, Kitzman, & Guyotte, 2006). One goal of this nonjudgmental observation is to gain the insight that all thoughts and feelings, regardless of their content, are empty and transient constructions. Nancy Hamilton, Heather Kitzman, and Stephanie Guyotte (2006) note that although these two approaches (CBT labeling versus mindfulness nonlabeling) are diametrically opposite in practice, they can lead to the same outcomes, and the two practices are frequently used in complement, as in MBCT.

Mindfulness: State or Trait?

Brown and Ryan argue that some people may have a disposition (i.e., trait) toward being more mindful, and so they have explored mindfulness as a naturally occurring attribute. Their research has shown that mindfulness as a trait has a positive effect on self-regulated activity and on wellbeing (Brown & Ryan, 2003; Brown, Ryan, & Creswell, 2007). However, they have also studied mindfulness as a state that can occur as a result of training, and their research has found that, independent of disposition, momentary experiences of mindfulness also have a salutary effect on wellbeing.

Research on mindfulness states, the success of programs such as MBSR, and the ancient history of Buddhist practice give ample evidence that mindfulness can be developed with practice, and therefore we can infer that there is potential for technology to be involved in this practice, be it as direct guidance or as peripheral support. But before we get to that, it is worth peeking into the world of the neuroscientist in order to discover how mindfulness physically changes our brain structures and their activity.

The Neuroscience of Mindfulness

Sitting in the confines of his plastic cave, the monk enters a state of deep, spacious awareness. Ready to detect any change in his brainwaves are the

24 electrodes pasted to his conveniently shaven head, and the fMRI machine—the cave within which he sits—is sending data on cerebral blood flow to the group of eager scientists behind the glass. Ancient in its origins, meditation is by far the oldest systematic practice for developing mindfulness that exists. Newfangled as mindfulness practice may sound to some, humans were training in it long before they were writing on paper. The practice has survived 2,000 years of human history, and modern-day Buddhist monks continue to engage in and teach these same practices around the globe. Happily, these same monks are also amenable to satisfying the empirical curiosity of neuroscientists, which is why we find them in unlikely places, such as fMRI machines in Madison, Wisconsin.

The physiological impacts of long-term meditation practice are surprisingly conspicuous from a neuroscientific point of view. Although early studies with expert meditators using EEG date back to the 1950s and 1960s, it's only during the past decade that we have been able to accumulate a more rigorous collection of neuroscientific evidence for the impacts of mindfulness practice.

Via EEG studies, mindfulness meditation has been associated with measurable changes to brainwaves, including changes in alpha waves and increases in theta and gamma rhythms (Lutz, Dunne, & Davidson, 2006). Alberto Chiesa and Alessandro Serretti (2010) report that, in addition to significant increases in alpha and theta activity, mindfulness meditation is associated with activation of the prefrontal cortex and anterior cingulate cortex, areas related to attention. Moreover, long-term meditation leads to enhancement of both these areas (specifically, the anterior cingulate cortex and the dorsolateral prefrontal). Less academically speaking, practicing mindfulness pumps up your attention muscles.

One surprising finding, with implications for those seeking to measure levels of mindfulness physiologically, is that although it seems logical to assume meditation practice will develop interoceptive awareness, which is often measured by a participant's ability to detect his own heartbeat, Sahib Khalsa and his colleagues (2008) have shown that practicing attention to internal body sensations does not actually lead to a better heart beat detection. So how does one go about measuring mindfulness? Well, usually, with a questionnaire …

Measuring Mindfulness

We can now use EEG and even brain imaging to detect and study mindfulness states, but the simplest, most widely used and thoroughly validated ways to measure mindfulness as both a state and trait are self-report instruments and established mindfulness scales developed by various research groups, each with a slightly different focus.

Brown and Ryan (2003) have provided a theoretical and empirical framework, including a dispositional scale (the Mindful Attention Awareness Scale, or MAAS) that can be used to measure the individual differences in the frequency of mindful states. MAAS consists of 15 statements to which participants assign a score from 1 to 6: for example, "I could be experiencing some emotion and not be conscious about it until some time later" or "I tend to walk quickly to get where I am going without paying attention to what I experience along the way." The questions have been confirmed by a number of studies that have indicated that the MAAS is a valid measure, distinct from other related measures.

The Freiburg Mindfulness Inventory is another widely used scale (Walach, Buchheld, Buttenmüller, Kleinknecht, & Schmidt, 2006) that was developed for use with experienced meditators but includes a later version adapted for use by nonmeditators. Other scales include the Kentucky Inventory of Mindfulness Skills, the Cognitive and Affective Mindfulness Scale–Revised, the Southampton Mindfulness Questionnaire, the Five Facet Mindfulness Questionnaire, the Philadelphia Mindfulness Scale, and the Toronto Mindfulness Scale. For more information on these scales, Ruth Baer (2011) provides a summary and evaluation of approaches to measuring mindfulness.

Fortunately, there is good correlation among scales as well as between scale measures and related psychological measures (and, to a less studied degree, there is correlation between the results of these scales and neurological measures). Nevertheless, researchers agree that available measures are never perfect, and further development to address weaknesses and devise complimentary methods of measurement is ongoing.

Measuring the Impact of Mindfulness on Wellbeing

The end goal for any effort in cultivating mindfulness is to foster wellbeing (whether by reducing the experience of pain, preventing depression, treating anxiety, or something else.) In order to measure the link between mindfulness and wellbeing, Brown & Ryan (2003) have used a combination of scales across several studies, including the CES-D Scale and the Beck Depression Inventory (for depression); the Positive and Negative Affect Scale and a scale of affective tone (for SWB); two scales to measure eudaimonic wellbeing; and two to measure physical wellness.

These studies showed that mindfulness (as measured by MAAS) was significantly correlated with measures of self-regulation and wellbeing. Furthermore, to address the limitations of retrospective self-reports, Brown and Ryan added another study using experience sampling (sending reminders to subjects via pagers). MAAS was used to assess both trait and state mindfulness, where "trait" referred to measures of intrapersonal differences, while "state mindfulness" measured variability of an individual's states. Both trait and state mindfulness were independent factors, and both had salutary effects.

Some of the limitations of Brown and Ryan's experience-sampling study (for example, the number of participants, the amount of data collected, the timing of the data collection) were addressed by a more recent study by Matthew Killingsworth and Daniel Gilbert (2010). They also used experience sampling, although thanks to technological advancements in the interim period, they were able to harness the convenience of mobile phones. In order to investigate whether mindfulness was a cause of happiness, they created a mobile phone application (see trackyourhappiness.org) that periodically interrupted participants during the day to ask the following three questions:

- *How are you feeling right now?* (answers were on a sliding 1–100 scale)
- *What are you doing right now?* (with a choice of 22 activities generally used in the day-reconstruction method (Kahneman, Krueger, Schkade, Schwarz, & Stone, 2004)
- *Are you thinking about something other than what you are doing?* (with the following options: "No," "Yes, something pleasant," "Yes, something unpleasant," or "Yes, something neutral")

The application collected an impressive 2.5 million samples from 5,000 individuals from 83 different countries ages 18 to 88. From among these samples, they analyzed the US-derived data ($N = 2,250$), and the findings were striking.

1. Mind-wandering occurred about half the time. Their findings suggest that on average 46.5 percent of the time we are thinking about something other than what we're doing. This happens at least 30 percent of the time for specific activities (with the exception of love making, for which we mind wander—or admit to it—only about 10 percent of the time). Interestingly, whether we mind wander is not, as one might expect, related to the nature of the activity we are engaged in.

2. People were less happy when their minds wandered, *regardless* of what they were doing at that time. Minds wandered to positive things 42.5

percent of the time, to negative things 26.5 percent of the time, and to neutral things 31 percent of the time. And here's the clincher: people were *not* happier when their minds wandered to positive thoughts than when they were simply being mindful of the task at hand (whatever it may be). In other words, fantasizing about a tropical island while sitting in dull traffic will not make you happier than mindfully attending to the traffic.
3. What people were *thinking* was more important to predicting how they felt than what they were *doing*. The nature of participant activity explained only 4.6 percent of the variance in happiness, while mind wandering explained 10.8 percent.
4. Most significantly, the study gave evidence that mind wandering was generally a *cause* rather than simply a consequence of unhappiness.

Clearly, mind wandering is a mental activity that sits in opposition to mindfulness in that it describes doing one thing but thinking another. It seems logical then to suspect that multitasking, which refers to the simultaneous execution of multiple tasks, might also reduce mindfulness and consequently wellbeing to some degree. Indeed, research is providing evidence for such a hypothesis.

In a recent survey of more than 3,000 girls ages 8–12 (Pea et al., 2012), media multitasking was shown to be negatively correlated to wellbeing (media multitasking constituted combinations of watching video, playing videogames, listening to music, texting, and talking over the phone). Specifically, the more participants multitasked, the worse they felt. In contrast, face-to-face social activities, which were also measured by the study, were positively correlated to wellbeing. Another study (Becker, Alzahabi, & Hopwood, 2013) measured 318 college students and found that increased media multitasking was associated with higher symptoms of depression and social anxiety (even after controlling for neuroticism, extraversion, and overall media use).

Multitasking research is particularly relevant to positive computing because the current design of technology actively facilitates parallel activity. This is probably because most digital devices are considered productivity tools, and multitasking is culturally viewed largely as a positive expression of greater productivity, despite ample evidence that it *reduces* performance (Wang & Tchernev, 2012).

Digitally afforded multitasking trains us to be drawn away from what we're experiencing and to be distractible. Whenever we come across a few seconds of pause (e.g., a file download, a line at the grocery store), we have come to react by immediately seeking something to fill in the moment, and mobile phones make this possible anywhere, anytime. Our brains seem addicted to busyness, despite our pleas for rest. Time and again we react to the prospect of present-moment awareness with anxiety, rapidly seeking out new points of attention—we even look for new thoughts to think (What can I plan? I should check my schedule mentally while I stand here) rather than settling into the moment as it is. Ironically, this is all at the expense of both performance and wellbeing.

With multiple devices on hand, each of which houses multiple applications displayed in parallel windows, modern devices are multitasking dynamos. It is only recently in response to information overload that some software designs have been switching some of the focus back to well ... focus. If multitasking does in fact decrease wellbeing, researchers in positive computing need to investigate how we can support productivity (genuinely) and in ways that don't compromise wellbeing.

The abundance of research evidence for the wellbeing benefits of mindfulness gives us plenty of reason to support it in the context of positive computing. However, how do we approach supporting users with a factor so subjective and internally experienced? Eight-week programs are unlikely to be the best model when users seldom engage with technology in the way they do with intensive medical interventions. Nevertheless, even longer programs are composed of smaller parts, and interventions in psychology remain an important place to look for guidance.

Strategies and Interventions for Fostering Mindfulness

Mindfulness Training and Meditation

Jon Kabat-Zinn's (1990) eight-week intensive MBSR program is based on a combination of meditation, Yoga practices, and inquiry exercises developed in weekly meetings. The meditation practice consists of turning attention, without judgment, to thoughts, feelings, and body sensations. A meta-analysis of 20 independent evaluations of MBSR programs (Grossman et al., 2004) found that all the interventions had similarly large effect sizes of 0.5 ($P < 0.0001$),[3] a track record of success that has placed MBSR programs in more than 200 institutions around the world.

Mindfulness-based training programs and mindfulness-based CBT programs incorporate multiple strategies, including mindfulness meditation. For mindfulness-based meditation (MBM in the scientific literature), various approaches have been used for guiding practitioners, including using mental practices such as body scans (intentionally observing and relaxing each part of the body) and the use of metaphors ("Simply watch your thoughts as you would watch clouds pass across the sky"). Buddhist teaching commonly employs attention to the breath as a technique for bringing a wandering mind back to the present (e.g., "Whenever you find your thoughts have wandered, simply bring your attention back to your breath, following it as it moves in and out"). The idea is not to encourage deep concentration on the breath, but simply to use the breath as a gentle and ever-available returning point.

The core of mindfulness meditation involves training the mind to reel itself back in whenever it wanders, as it wanders, over and over again. As Pema Chödrön (2007) puts it, "Through meditation, you're training in interrupting the momentum of the wandering mind and going right to the experience itself." As we have come to learn from discoveries on neuroplasticity, the mind can indeed be trained and changed. We can get better at this.

Although most modern technology trains us in splitting attention, there are a number of initiatives, from the experimental to the commercial, that have already been developed specifically to support mindfulness, either by providing guidance for meditation or by training the wandering mind to return to the present through the use of sound or visual stimuli (we look at those technologies in the later section "Digital Technologies for Mindfulness").

Strategies in Education

Despite the fact that William James had already espoused the value of mindfulness training for education back in the nineteenth century, we are only now beginning to see this training find its way into the curriculum. Politicians, academic administrators, and educators are gradually coming to the conclusion that promoting mindfulness programs in schools is worth serious attention.

Tim Ryan (2012), a Democratic congressman from Ohio, has shared his views on the ways in which mindfulness can change schooling, the healthcare system, the military, and even the nation. His basic tenet is that mindfulness is an important aspect of the socioemotional skill set that leads to good conflict resolution, responsible decision making, better relationships, the setting of goals, and the development of self-discipline. Together with Judy Biggert, then Republican congresswoman from Illinois, Ryan sponsored legislation to introduce socioemotional learning in schools, which included the Academic, Social, and Emotional Learning Act of 2011.

The Association for Mindfulness in Education, a collection of organizations and individuals promoting the introduction of mindfulness practices in education maintains a list of schools and programs that already include mindfulness in their curriculum.[4]

Integrating mindfulness training programs or other mindfulness interventions into people's lives through schooling and workplace training requires the commitment of policymakers and managers who recognize the value of such programs. Seeing that scientists have shown the causal relationship between mindfulness and wellbeing, it seems rational to expect an increasing number of initiatives to emerge over the next decade.

Biofeedback Interventions

As mentioned earlier, there are measurable correlations between mindfulness states and various physiological signals (Chiesa & Serretti, 2010; Lutz et al., 2006), and although these signals are useful for research, some researchers have gone further to ask how they might be used as feedback for mindfulness training. Biofeedback systems record physiological signals and feed them back to the user as sound or visual stimuli in real time. The signals may include a variety of inputs from EEG and electrocardiogram to heart rate or breathing patterns. By receiving this synced real-time feedback, a participant can learn to change a detected physiological factor, for example, via operant conditioning (with a subtle reward system).

There is some minimal early evidence that biofeedback systems could be helpful in supporting mindfulness or at least some of the characteristics associated with mindfulness, such as stress reduction, emotional awareness, mental clarity, and loss of intention (Plasier, Bulut, & Aarts, 2011; Stinson & Arthur, 2013; Vidyarthi & Riecke, 2013). This early work suggests that biofeedback systems may provide a useful area of future research and development.

Although the price, size, and accessibility of biofeedback sensors and systems have been plummeting over the past few years, most experimental systems remain fairly intrusive or expensive or both. An alternative approach comprises a challenge to positive computing: How can mindfulness states and training be effectively supported by common digital technologies?

Digital Technology for Mindfulness

From melting into an embodied symphonic experience to following the reassuring guidance of a soft voice recording or even sharing your efforts with friends, the examples of technology-mediated support for mindfulness, though in their early stages, are already surprisingly diverse. Available technologies designed around mindfulness practices can currently be placed into three categories: (1) those focused on guided sessions (lessons, exercises, or meditation sessions); (2) those that include social networking and sharing features; and (3) those focused on embodied experiences, often based on biofeedback approaches. Numerous applications have been designed to promote mindfulness, but we have selected just a few to describe by way of example.

Guided Sessions

Smiling Mind is a collection of mindfulness-training activities delivered over the Internet and via mobile app. Like other audio-instruction-based courses, it provides a series of guided meditation recordings. It differentiates itself, however, by providing these recordings in the context of tailored training curricula. Smiling Mind provides content targeted to various contexts (e.g., a curriculum program aimed at school-age children and a corporate training program for adults). For each of the programs, the focus has been on developing a high-quality curriculum and set of materials.

Other examples of guided support have specifically targeted stress in the workplace. Mark Williams and his group at the University of Oxford (Krusche, Cyhlarova, King, & Williams, 2012) evaluated a set of modules from the MBSR and MBCT programs delivered over six weeks. The Perceived Stress Scale (Cohen, Kamarck, & Mermelstein, 1983) was used to measure the program's impact, and the study concluded that online mindfulness training can significantly decrease perceived stress (with changes still apparent at a one-month follow-up).

Social Features in Mindfulness Training

Projects such as Mindfulnets[5] distinguish themselves by adding social networking features that allow users to share their ongoing experience with others. The user fills out two questionnaires, the Perceived Stress Scale and the Five Facets Mindfulness Questionnaire. Although the responses from the questionnaires are not used to personalize the interaction (which might be an interesting feature in future), they are used to measure the impact of the interventions, which is useful both for the user (as evidence of improvement) and for the organization (as evidence of efficacy).

An interesting feature of Mindfulnets is its integration with Facebook. Users are able to comment on their progress, and discussions are visible to visitors. Another feature is a page displaying personal statistics and results of the Perceived Stress Scale and the Five Facets Mindfulness Questionnaire. These sharing features could hypothetically have two types of impact. On the one hand, the sharing of experience may increase motivation. On the other hand, users may be more likely to compare themselves to others and judge themselves, which is recognized as a disabling practice for mindfulness. The website intervention was evaluated (Quintana & Rivera, 2012) giving small effect sizes, albeit with a number of disclaimers due to the small and sparse distribution of the subjects' demographics.

One study (Morledge et al., 2013) evaluated a mindfulness intervention for which a content-centric program was compared to the same intervention but combined with discussion forums. The results reproduced the positive effects of mindfulness intervention shown in other studies, with statistically significant improvements in perceived stress and mindfulness for both the content-based version and the version with social features at 8 and 12 weeks. With regard to the social features, the authors stated, "Some evidence from our study suggested that this component improved some participants' therapeutic experience. These results and feedback from participants suggest that greater benefits may be achieved with a more expansive and integrated social media component."

More research is currently needed before we can understand if social interaction and sharing are beneficial to mindfulness practice and interventions, which are by nature personal endeavors. If so, we will also need to know what kinds of social features are helpful (and which are harmful), for whom, and in what contexts, and what design features have an impact.

Embodied Experience

Very few have explored embodied experiences for supporting mindfulness, but notable exceptions do exist. At a basic level, the mindfulness breathing exercises at Mindfulnets include very simple embodiment features in that the user clicks the mouse with each breath while a preset timer runs.

For a more thoroughly embodied experience, we turn to Sonic Cradle. In the Sonic Cradle, you are suspended in a hammock-like "cradle" while straps fastened around your chest detect your breathing patterns. These breathing patterns are then used to shape the ambient sound enveloping you in the dark sound-proof space. Created by HCI designer and musician

Jay Vidyarthi, Sonic Cradle is an immersive full-body experience designed to mediate mindfulness meditation, particularly as a way to introduce the practice to novices (Vidyarthi & Riecke, 2013).[6] A qualitative investigation found that the experience led participants to experience subjective elements of mindfulness meditation—in particular, clarity of mind and loss of intention.

But mindful awareness is by no means restricted to the meditative context and is intended as a way of life. Buddhist monks practice mindful walking, eating, cleaning, and living in general. Mindful eating is gaining popularity in the mainstream, and programs have even found their way into corporate offices (Google recently invited Zen teacher Thich Nhat Hanh to come and give their employees guidance on mindful eating).

A few technologies already tackle mindfulness support in day-to-day activity. For example, the Breath–Walk Aware System was designed to support beginning meditators in the practice of walking meditation using a smartphone (Yu, Wu, Lee, & Hung, 2012). Similarly, the Slow Floor employs pressure-sensitive surfaces and sound to promote awareness of bodily movement (Feltham & Loke, 2012).

The HAPIfork is a commercial product that intervenes at mealtimes by vibrating when you eat too fast. The electronic fork promises to promote mindful eating (and associated weight loss) and is also a data-collection tool that records the time it took you to finish a meal and the speed at which you ate (measured by "fork servings" per minute and intervals between them). It will also connect with apps to correlate your eating data with sleep, meal, and relaxation times. The fork's associated desktop app has coaching advice and provides stats from data collected automatically by the fork and self-reports.

From light-up floors to shaking forks, we're clearly only seeing the very tip of the iceberg when it comes to how embodiment might be employed to support mindful attention during meditation and daily activity. It will be a while yet before we can come to know what is genuinely effective and what is obstructive or merely superfluous to such endeavors, and we can expect there to be differences across contexts, levels of experience, and possibly cultures. At any rate, embodiment research will be an interesting space to watch (and sample) as future developments arise.

Games for Mindfulness?

Although there are websites eager to suggest that common videogames train us in mindfulness, they are confusing mindfulness with concentration. We can concentrate heavily on a videogame for hours, *losing our awareness* of what is happening around us. We do this in the context of *striving* to achieve a specific goal. Concentrated striving is not mindfulness. Mindfulness is frequently described as a method of "nonstriving" (see Williams & Kabat-Zinn, 2013) that facilitates open acceptance and awareness, so goal setting seems decidedly antithetical in this context. This indeed poses challenges for the use of games for mindfulness. Can you have a game without an object or goal? Would it still be a game by definition? For this reason, we may find that games are better suited to training certain skills that are helpful to mindfulness (e.g. returning to the breath) than to inducing a state of mindfulness itself. In fact, this is precisely the route being taken by the earliest experiments in mindfulness games.

Games+Learning+Society (GLS), a group of researchers and developers based in Madison, Wisconsin, is working on games specifically designed to provide practice in aspects of mindfulness and other wellbeing-related skills. One of these games, called *Tenacity*, requires the player to tap an iPad once for each breath and twice for each fifth breath through increasing levels of distracting difficulty. Researchers at GLS, who include neuroscientists such as Richard Davidson, study the psychological effects of videogames using a combination of behavioral, structural, and fMRI-based measures—an approach that is sure to reveal important insights for positive computing designers.

Design Implications

Distraction versus Guidance

In Japanese monasteries, the head monk voluntarily provides a generous service to those in his spiritual care—he walks around the hall quietly until he spots someone whose mind has evidently wandered, at which point he raps the willing student on the shoulder with a stick. Veterans swear by the effectiveness of this practice. Still, some prefer the comparatively gentle awakening provided by the meditation bell or the early-morning gong. Each of these methods, from the stick to the gong, is a form of short and sweet wake-up call designed to bring a wandering mind back to the present moment, and there doesn't seem to be any reason that technology couldn't be employed to provide a similar service. Yet how do we separate the song of the bell from the nag of the beep? A beep is an aggravation if it goes off at the wrong time, and a rap with a stick unsolicited could lead to bursts of profanity. At least part of the answer to striking this balance seems to lie in two things: *autonomy* (as usual) and that design stalwart, *minimalism*.

First, in the case of autonomy, whether the stick is an affront or an invigorating aid has something to do with whether you agreed to it or not. For software design, that means honoring user agency as to when, where, and how help occurs. Moreover, autonomy issues can extend into important details. Can the user set limits, apply variations, or otherwise take part in shaping the support environment or service? After all, it's ultimately a kind of improved autonomy that one is attempting to foster with mindfulness training, in the form of improved self-regulation of attention.

Second, there is reason to believe that minimalism is critical to design intended to support mindfulness. We have already discussed how modern digital environments, with their continuous provision of distraction, train us in splitting attention. Meditation halls and retreat environments are effective for mind training precisely because they are comparatively distraction free. Therefore, introducing additional distraction is counterproductive to supporting a more conducive environment for practice. As we improve in our capacity for mindfulness we should of course be able to practice it even amidst distraction and difficulty, but few of us start at that level. Of course, a notable exception is when a particular training program introduces distraction deliberately as part of training participants in bringing focus back to the present (as with *Tenacity*).

By minimalism we also mean subtlety. You might say the Zen master's stick is hardly subtle, but then it is simple, direct, infrequently applied, and reserved for willing experts. In general, traditional methods such as gongs and bells are applied in a way that is infrequent, gradually increases in volume, and is prefaced and followed by silence. The intention is not to devastatingly startle or distract, but to gently awaken from mind-wandering reverie.

Using Aural and Haptic Feedback

The bell and stick go back centuries as effective tools for supporting mindfulness and meditation. In modern terms, they might be described as tools for aural and haptic feedback. Another form of haptic feedback is seen in a common meditation posture that requires holding the hands in the lap with thumbs *almost* touching (the "Zen mudra") (Chödrön, 2007). This clever feedback mechanism is designed to let you know you're dozing off (as your thumbs collapse) or overdoing it (as they push together).

Some of the modern examples mentioned previously, such as Sonic Cradle, have successfully employed aural feedback to represent physiological signals. In another example, the Tibetan "singing bowl" inspired an experimental variation: the Electronic Singing Bowl produced by a team at Phillips (Plasier et al., 2011). The bowl produced synthesized gong sounds and monaural beats (sound-frequency combinations that have been shown to trigger states of relaxation). The device was not designed to support mindfulness, but rather to aid relaxation, for which it was shown to be helpful. Even here, however, a design principle of minimalism is advisable as participant feedback from such studies shows that users have an aversion to sounds that are unusual, overly loud, or overly present—probably because meditation is something generally done in silence.

Supporting Nonjudgment
We have already alluded to the potential pitfalls of applying certain motivational features such as tracking and goal-setting to meditative practice and mindfulness training, although they are highly effective in other contexts. We also highlighted the "nonstriving" aspect of mindfulness that makes goal setting problematic. Tracking and goal setting may also encourage comparison to others ("I will practice mindfulness every day this month" or "I can't believe he meditates more than me").

Jon Kabat-Zinn (2003) explains that in MBSR clients are encouraged "to let go of their expectations, goals, and aspirations for coming, even though they are very real and valid, to let go—momentarily, at least … and to simply 'drop in' on the actuality of their lived experience and then to sustain it as best they can moment by moment, with intentional openhearted presence and suspension of judgment and distraction, to whatever degree possible." Thus, we may find that for supporting mindfulness it may be the relinquishing of goal setting, expectation, and striving that requires the most support.

Yet how do you reconcile the need for purpose and direction in lifestyle change as well as people's motivations for pursuing practices like mindfulness with the essential attitude of nonstriving? Kabat-Zinn encourages participants to let go "momentarily—at least." The idea is not to chuck away all your goals in life, but to let go of striving, tallying, and resistance during mindfulness practice. Buddhist and Yogic practices find a balance between goals and nonstriving by encouraging participants to "make aspirations" or "set intentions for the day." In this broader approach, aspirations provide direction but are sufficiently vague and gently applied as to be easily let go. More specifically, they are not quantified.

In other words, an intention may be "I focus today on returning to the present moment each time I realize my mind has wandered" and less like

"I will clock 20 hours of mindfulness meditation this month." Also, individuals set their own intentions; they are not supplied with a prewritten structure of achievement levels by someone else.

Support for self-compassion may also prove helpful in balancing out our tendency to strive and judge in this context. Kabat-Zinn (2003) suggests the answer to reconciling nonstriving with the reality of valid intentions is rooted in teacher authenticity, which in our case, translates to designer authenticity.

Practicing What You Teach (or Design For)

Although there do not seem to be any downsides to mindfulness (for example, there are no reported negative side effects), Kabat-Zinn has drawn attention to the potential pitfalls that scientists (and, we infer, technologists) are at risk of encountering as interest in mindfulness gains momentum within clinical practice and beyond. In a thorough commentary, Kabat-Zinn (2003) cautions:

> It becomes critically important that those persons coming to the field with professional interest and enthusiasm recognize the unique qualities and characteristics of mindfulness as a meditative practice, with all that implies, so that mindfulness is not simply seized upon as the next promising cognitive behavioral technique or exercise, decontextualized, and "plugged" into a behaviorist paradigm with the aim of driving desirable change, or of fixing what is broken.

Specifically, he challenges us to practice mindfulness systematically, *before* we attempt to teach it or support its development in others. He warns us not to ask more of our patients (or users) than we do of ourselves.

If we fail to practice first, either because we are too busy or not sufficiently interested, Kabat-Zinn (2003) warns that our "attempts at mindfulness-based intervention run the risk of becoming caricatures of mindfulness, missing the radical, transformational essence and becoming caught, perhaps by important but not necessarily fundamental and often only superficial, similarities between mindfulness practices and relaxation strategies, cognitive-behavioral exercises, and self-monitoring tasks."

In the case of technology, we can just as easily get carried away by enthusiasm, by our desire to fit a mindfulness agenda into a particular technology or vice versa, or by a temptation to compromise critical principles of mindfulness for the sake of quantifiable goals or aesthetic values. The lesson is simple: when it comes to mindfulness, don't just talk about it, don't just design for it, do it, and do it before you design.

Expert Perspectives—Technology for Mindfulness

Mindfulness Online

Adele Krusche and J. Mark G. Williams, Oxford Mindfulness Centre, University of Oxford

Computing is not about computers any more. It is about living.
—Nicholas Negroponte, *Being Digital*

At the Oxford Mindfulness Centre, we teach people the skills to be able to let go of unwanted thoughts, to sit still and to pay attention to what is going on right here and now. Within Oxford University's Department of Psychiatry, we work with partners around the world to prevent depression and enhance human potential through the therapeutic use of mindfulness. The benefits of mindfulness in preventing serious depression and emotional distress have been demonstrated by clinical trials, and work continues to explore the implications of this relationship and to extend the approach to other disorders, using brain-imaging techniques together with experimental cognitive science to learn how mindfulness has its effects and which practices are best for whom.

The society we live in and mindfulness therapy seem to be poles apart. A simple concept such as staying still and noticing seems, to many, very alien now, and perhaps that means we need to remember it. Mindfulness courses delivered in person have been shown to be effective for many people experiencing a variety of problems.

(continued)

> The problem comes when we have classes on mindfulness running a few miles from home, starting perhaps while we are at work or picking the kids up from school; maybe disability or cost gets in the way. This is where online mindfulness comes in. It offers the opportunity for people to learn the skills in their own homes, in their own comfortable chair, without the need to rush or arrange childcare. These skills can then be used in the very surroundings they might need most.
>
> That isn't to say that this online mindfulness therapy is for everyone. We're only just starting to explore its effectiveness. Our preliminary research suggests that Internet-based mindfulness training has the potential to reduce stress, anxiety, and depression in people who have chosen to take a course. We don't yet know much about the people taking the course, but are looking forward to finding out exactly who and who does not benefit from it. If people are getting results from this course equivalent to online CBT or some face-to-face courses, it seems essential to continue the research and to develop online therapies to help the people who choose to take them. The word is starting to spread that these Internet interventions exist, that helpful therapies are available now, and that they might just make you feel better without your having to leave your living room. Being able to steal some time back in the light of your monitor might take some of the frantic out of the day.
>
> For further reading, see Krusche et al. 2012. For mindfulness online, a preliminary evaluation of the feasibility of a web-based mindfulness course and the impact on stress is available at *BMJ Open, 2*(3).

Notes

1. The eight steps are: Right View, Right Intentions, Right Speech, Right Actions, Right Livelihood, Right Effort, Right Mindfulness, and Right Concentration.
2. The research guide can be found at mindfulexperience.org.
3. This means that the relationship between the intervention (doing MBSR) and wellbeing is considered high.
4. For this promotion, see the website mindfuleducation.org.
5. See the website mindfulnets.co.
6. Although the full experience requires the cradle and a sound-proof room, you can get a taste of it at jayvidyarthi.com.

References

Baer, R. A. (2011). Measuring mindfulness. *Contemporary Buddhism, 12*(1), 241–261.

Becker, M. W., Alzahabi, R., & Hopwood, C. J. (2013). Media multitasking is associated with symptoms of depression and social anxiety. *Cyberpsychology, Behavior, and Social Networking, 16*(2), 132–135.

Brown, K. W., & Ryan, R. M. (2003). The benefits of being present: Mindfulness and its role in psychological well-being. *Journal of Personality and Social Psychology, 84*(4), 822–848.

Brown, K. W., Ryan, R. M., & Creswell, J. D. (2007). Mindfulness: Theoretical foundations and evidence for its salutary effects. *Psychological Inquiry, 18*(4), 211–237.

Chiesa, A., & Serretti, A. (2010). A systematic review of neurobiological and clinical features of mindfulness meditations. *Psychological Medicine, 40*(8), 1239–1252.

Chödrön, P. (2007). *How to meditate with Pema Chodron: A practical guide to making friends with your mind.* Boulder, CO: Sounds True.

Cohen, S., Kamarck, T., & Mermelstein, R. (1983). A global measure of perceived stress. *Journal of Health and Social Behavior, 24*(4), 385–396.

Feltham, F., & Loke, L. (2012). The slow floor: Towards an awareness of bodily movement through interactive walking surfaces. Paper presented at the workshop "The Body in Design," at the 24th Australian Computer-Human Interaction Conference, OzCHI '12, Melbourne, VIC, Australia, November 26–30.

Grossman, P., Niemann, L., Schmidt, S., & Walach, H. (2004). Mindfulness-based stress reduction and health benefits. A meta-analysis. *Journal of Psychosomatic Research, 57*(1), 35–43.

Hamilton, N. A., Kitzman, H., & Guyotte, S. (2006). Enhancing health and emotion: Mindfulness as a missing link between cognitive therapy and positive psychology. *Journal of Cognitive Psychotherapy*, *20*(2), 123–134.

Hanh, T. N. (2008). *The miracle of mindfulness*. New York: Random House.

James, W. (1892). *Psychology: A briefer course*. New York: Henry Holt.

Kabat-Zinn, J. (1990). *Full catastrophe living: Using the wisdom of your body and mind to face stress, pain, and illness*. New York: Delacourt.

Kabat-Zinn, J. (2003). Mindfulness-based interventions in context: Past, present, and future. *Clinical Psychology: Science and Practice*, *10*(2), 144–156.

Kahneman, D. (2013). *Thinking, fast and slow* (p. 512). New York: Farrar, Straus and Giroux.

Kahneman, D., Krueger, A. B., Schkade, D. A., Schwarz, N. S., & Stone, A. A. (2004). A survey method for characterizing daily life experience: The day reconstruction method. *Science*, *306*(5702), 1776–1780.

Khalsa, S. S., Rudrauf, D., Damasio, A. R., Davidson, R. J., Lutz, A., & Tranel, D. (2008). Interoceptive awareness in experienced meditators. *Psychophysiology*, *45*(4), 671–677.

Killingsworth, M. A., & Gilbert, D. T. (2010). A wandering mind is an unhappy mind. *Science*, *330*(6006), 932.

Krusche, A., Cyhlarova, E., King, S., & Williams, J. M. G. (2012). Mindfulness online: A preliminary evaluation of the feasibility of a web-based mindfulness course and the impact on stress. *BMJ Open*, *2*(3). Retrieved from http://bmjopen.bmj.com/content/2/3/e000803.full.html.

Lutz, A., Dunne, J. D., & Davidson, R. J. (2006). Meditation and the neuroscience of consciousness: An introduction. In P. D. Zelazo, M. Moscovitch, and E. Thompson (Eds.), *The Cambridge Handbook of Consciousness* (pp. 497–550). Cambridge, UK: Cambridge University Press.

Morledge, T. J., Allexandre, D., Fox, E., Fu, A. Z., Higashi, M. K., Kruzikas, D. T., ... Reese, P. R. (2013). Feasibility of an online mindfulness program for stress management: A randomized, controlled trial. *Annals of Behavioral Medicine*, *46*(2):137–148.

Negroponte, N. 1995. *Being digital*. New York: Vintage.

Pea, R., Nass, C., Meheula, L., Rance, M., Kumar, A., Bamford, H., ... Zhou, M. (2012). Media use, face-to-face communication, media multitasking, and social well-being among 8- to 12-year-old girls. *Developmental Psychology*, *48*(2), 327–336.

Plasier, S., Bulut, M., & Aarts, R. (2011). A study of monaural beat effects on brain activity using an electronic singing bowl. Annual Symposium of the IEEE

Engineering in Medicine and Biology Society (EMBS). Benelux Chapter, December 2 (Brussels/Leuven, Belgium).

Quintana, M., & Rivera, O. (2012). Mindfulness training online for stress reduction, a global measure. *Studies in Health Technology and Informatics, 181*, 143–148.

Ryan, T. (2012). *A mindful nation*. Carlsbad, CA: Hay House.

Segal, Z. V., Williams, J. M. G., & Teasdale, J. D. (2012). *Mindfulness-based cognitive therapy for depression* (p. 471). New York: Guilford Press.

Speca, M., Carlson, L. E., Goodey, E., & Angen, M. (2000). A randomized, wait-list controlled clinical trial: The effect of a mindfulness meditation-based stress reduction program on mood and symptoms of stress in cancer outpatients. *Psychosomatic Medicine, 62*(5), 613–622.

Stinson, B., & Arthur, D. (2013). A novel EEG for alpha brain state training, neurobiofeedback, and behavior change. *Complementary Therapies in Clinical Practice, 19*(3), 114–118.

Vidyarthi, J., & Riecke, B. (2013). Mediated meditation: Cultivating mindfulness with sonic cradle. In *CHI13 extended abstracts on human factors in computing systems* (pp. 2305–2314). New York: ACM.

Walach, H., Buchheld, N., Buttenmüller, V., Kleinknecht, N., & Schmidt, S. (2006). Measuring mindfulness—the Freiburg Mindfulness Inventory (FMI). *Personality and Individual Differences, 40*(8), 1543–1555.

Wang, Z., & Tchernev, J. M. (2012). The "myth" of media multitasking: Reciprocal dynamics of media multitasking, personal needs, and gratifications. *Journal of Communication, 62*(3), 493–513.

Williams, J. M. G., & Kabat-Zinn, J. (Eds.). (2013). *Mindfulness: Diverse perspectives on its meaning, origins, and multiple applications at the intersection of science and dharma*. New York: Routledge.

Yu, M.-C., Wu, H., Lee, M.-S., & Hung, Y.-P. (2012). Multimedia-assisted breathwalk-aware system. *IEEE Transactions on Bio-medical Engineering, 59*(12), 3276–3282.

10 Empathy

Milo is a small gray mouse. His first memory is of feeding from his mother in a small cage, remarkably similar to the one he's in now. For as long as he can remember, the man in the white coat has visited periodically with pellets, toys, electrodes, and other variations to his otherwise solitary and monotonous existence. Two weeks ago Milo's life changed when he got the unexpected delight of a cagemate, an affectionate white mouse named Lula. They play together, chase each other around the cage, and sometimes even pile up in the corner for warmth (lab air conditioning is absolutely polar). Life is better with Lula, but Milo has found that when she hurts, so does he. One day, the man in the white coat came over, opened the cage door, and pinched Lula on the back of the neck with something sharp …

With the use of names, subjectivity, indicators of similarity, and signs of affection—the help of just a few simple social cues—this summary of a lab experiment becomes a story with characters with whom we can empathize. In the lab, Milo was probably listed as mouse 456b for the same reason farmers don't name their pigs. But the story of Milo holds more than one clue to empathy. Dale Langford and his colleagues (Langford, 2010; Langford et al., 2006) tested the interactions between pain sensitivity and mouse behavior (for strangers versus cagemates), revealing groundbreaking evidence of empathy in mice. Specifically, their work has shown that pain among mice is shared (viewing another mouse in pain increases a mouse's own sensitivity to pain), that mice express pain through recognizable facial expressions, that female mice will approach their cagemates when they see they are in pain, and that this may even provide comfort ("analgesic effects") (Langford, 2010).

Evidence of empathy in animals as dissimilar to humans as mice may or may not be surprising depending on your view of the animal kingdom. Many of us generally think of empathy as being something exclusively human, but then we tend to underestimate other animals with regularity, placing *Homo sapiens* on a rather elaborate pedestal. Of course, this anthropocentric view may simply be a by-product of necessary pragmatism. The fact that we modulate our empathy for animals is unsurprising because it is necessary if we are to use them for food, clothing, and scientific experiment as much as we do—all of which would be decidedly less palatable if our empathy for them was more intensely felt. By comparison, most people (barring a few cultural exceptions) are loath to the idea of eating dogs. This is not because dogs are more intelligent than the pigs we do eat (pigs are strikingly intelligent), but simply because so many of us have formed empathic relationships with domesticated dogs. They are part of the human family. We bear daily witness to the many ways in which dogs are similar to us—we see their joy, their fear, their loving attention—and we feel for them.

People's capacity for empathy has been influenced in many ways throughout history. It has been deliberately eroded through the use of stereotyping and depersonalization, as in Nazi propaganda. It has been fostered by literature, art, and community-building efforts. Empathy is frequently thought of in terms of sharing the negative (misery loves company), but it is equally valid for the sharing of positive emotions. Empathy is a development of animalian evolution with distinct advantages (de Vignemont & Singer, 2006), and its various neurological foundations are beginning to be revealed.

Empathy is a fascinating aspect of being human that is essential to healthy relationships, to collaboration, to wellbeing, and to personal growth. Remarkably, some inspiring technologies, from work with autistic children to virtual-reality games, have been designed specifically to develop it. It is for these reasons that we dedicate a chapter to empathy, to its scientific grounding, and to some of the many strategies for fostering it employed by psychology, art, and technology.

Understanding Empathy

Empathy is an essential component of human communication. By vicariously experiencing what another feels, we can understand their experience in a way that goes beyond the cognitive processing of linguistic or metalinguistic expressions. Consensus on a precise scientific definition of empathy is surprisingly elusive, but most variations reveal empathy to be a multifaceted construct that includes emotion recognition, vicarious feeling, and perspective taking (Singer, 2006). If empathy were to manifest as a profession, it might show up as social work, and the *Social Work Dictionary* defines empathy as "the act of perceiving, understanding, experiencing, and responding to the emotional state and ideas of another person" (Barker, 2008, cited in Gerdes, Segal, & Lietz, 2010).

The social work definition reveals there are both cognitive and affective sides of empathy. *Cognitive empathy* (often referred to as "theory of mind") is the ability to *recognize* emotions and intentions in others. *Affective empathy* is our ability to *share* the feelings of others and to react with appropriate emotion to what someone else is feeling or thinking. The distinction is physically visible in the brain in that each of these aspects relies on separate neuronal circuitry (Singer, 2006), and, as such, each can be developed independently of the other. For example, the sociopath may have well-developed cognitive empathy such that he is able to effectively lie, persuade, and make friends, but without affective empathy he fails to react with appropriate emotions such as guilt or remorse, which often leads to violent behavior. In contrast, those with autism spectrum disorders may have poor cognitive empathy, making it very difficult for them to discern what others are feeling and thinking, but they will be visibly distressed when others are suffering.

But a two-sided model is not quite enough. The recent accumulation of evidence for distinct neural processes associated with various aspects of empathy has begun to lend a level of concreteness to our understanding of it. Based on a combination of psychology and social neuroscience research, Jean Decety and Yoshiya Moriguchi (2007) propose a model of empathy that has four components:

1. **Affective sharing** between the self and the other.
2. **Self-awareness**, which prevents confusion between the self and other.
3. **Mental flexibility**, which allows the adoption of the subjective perspective of the other (perspective taking).
4. **Regulatory processes** that modulate the feelings associated with emotion.

A cross-disciplinary academic definition of empathy awaits us in the future, but more critical to positive computing is whether empathy is something fixed or something we can actually develop, and if so, how.

Developing Empathy

Evidence implicates both nature and nurture in the process of developing empathy in human beings, with impact factors as far ranging as temperament and genetics to parental style, early life experiences, pet ownership, synchronous movement, and meditative practices. One thing is clear, empathy is something that most certainly can be developed (for example via school-based intervention programs or even digital experiences).

In the now iconic psychological experiment on obedience, Stanley Milgram (1963) encouraged subjects to inflict strangers with electric shocks in order to investigate how far obedience to perceived authority could influence people to act in contradiction to their conscience and override feelings of empathy. His intention was to understand how so many ordinary people could have engaged actively in the atrocities of the Holocaust during the Second World War.

Although the impact of war, charismatic leadership, and propaganda are all critical parts of ongoing empathy research, it is from the research done within the context of ordinary life that positive computing has most to learn. For example, the power of family and school relationships to foster or inhibit the growth of empathy is immense. Parental attention and interaction in early childhood have been shown to be critical to the development of both the cognitive and affective sides of empathy (Farrant, Devine, Maybery, & Fletcher, 2012). Not only does a lack of parental empathy unsurprisingly predict neglectful parenting, but neglectful parenting leads to underdeveloped empathy in children—an unfortunate cycle, the consequences of which reverberate throughout society. As such, some measures for the prevention of child abuse, crime, and bullying approach these problems via programs for developing empathy in schools and among at-risk adolescents.

Empathy and Art

One of the most compelling arguments for the critical importance of the arts is their singular ability to expand our perspective, allow us to experience other lives, and deepen our sense of empathy and compassion. Whether it is through witnessing atrocities in a movie such as *Schindler's List*, identifying with the feelings of loss expressed in the essential lines of an Auden poem, being transported to the streets of Victorian England through the portal of a Dickens novel, opening to the emotional pangs brought forth by a Munch painting, or savoring the rare honesty and beauty of an Otis Redding song, great art has an unparalleled capacity to cut across boundaries of time, space, and language, to allow experiences otherwise inaccessible, to explore a shared humanity, and to foster personal growth. Great art will challenge us by helping us to understand not only the suffering of the oppressed, but also the suffering of the oppressors. The popularity of work by writers such as Shakespeare persists because even his villains are complex and human—they have something to teach us, something with which we can identify—hints at the precursors of immorality, the seeds of which are in everyone (as Milgram showed). In this way, art is one of humanity's greatest allies for understanding ourselves and others, and art's greatest ally is probably empathy.

In general, the ways in which technologies might someday foster empathy will likely be very different to the emotionally direct and often nonverbal ways in which art so often does. However, there is one area in which we may find that there are some overlaps and convergences, and that is amid the narrative and role-playing potential of digital games. We get to that later in this chapter, but, first, let's lighten the mood.

Empathic Joy

One utterly underdiscussed aspect of empathy, in Western culture at least, is empathic joy. Empathy is generally associated with sharing in another's suffering, and the German word *schadenfreude* has been appropriated into a number of languages to allow us to communicate the notion of "enjoying another's suffering" in one word. Yet we have somehow escaped the need to find a good word for "enjoying another's happiness" (although the Yiddish word *naches* makes occasional appearance and refers to the pleasure one takes in one's children and grandchildren).

Despite our general linguistic failure to represent a state of empathic joy, we frequently do experience it. We share in the natural jubilance of children; we are overjoyed at hearing a brother has landed a great job or a best friend has found love. We cry for joy at weddings and births and smile uncontrollably at the laughter of others.

Empathic joy is given considerably more attention in Buddhist psychology, and *mudita*, a Sanskrit word meaning "sympathetic or vicarious joy," is listed as one of four sublime states of mind. Together with loving kindness, equanimity, and compassion, vicarious joy is considered an ideal mind state for optimal social wellbeing. According to Buddhist teacher, Ven. Nyanaponika Thera (1999), these states "level social barriers, build harmonious communities, awaken slumbering magnanimity long forgotten, revive joy and hope long abandoned, and promote human brotherhood against the forces of egotism." Could technology be designed to foster *mudita*? It just might need to if recent studies are correct in indicating that we are losing our sense of empathy.

Is Empathy Taking a Downturn?

As mentioned in the chapter on self-awareness, birth cohort studies have shown evidence for changes in dispositional narcissism and specifically in empathy over the past 30 years. A cross-temporal meta-analysis of American college students (Konrath, O'Brien, & Hsing, 2011) showed a sharp decline in standard measures of empathic concern, and, most notably for positive computing, this decline was particularly sharp from the year 2000 on. Alignment of the trend with growth in new media has led the authors to speculate a relationship between the two.

The apparent rise in narcissism and decrease in empathy is still hotly debated, and although the causes for such changes may not be fully understood, the possible relationship with an increasingly digital environment is difficult to ignore and certainly worth confronting from a positive-computing perspective. There are pockets of further evidence. For example, as with other forms of violent media, there is significant evidence that violent videogames decrease empathy and physiologically desensitize users to violence (Anderson et al., 2010), but then prosocial games increase prosocial behavior (Gentile et al., 2009). This is critical to positive computing because gaming makes up an ever-increasing chunk of our digital experience, and it makes more sense to look at videogames' psychological impacts from a higher level and holistic angle rather than inadvertently suggesting their influences go only one way. To that end, Katherine Buckley and Craig Anderson (2006) provide a learning-based theoretical model to help organize and explain the full gamut of positive and negative effects of videogames.

Social media likewise have given us reason to believe that technology, through facilitating either cyberbullying or shared experience, can detract *or* support empathy in many ways. What's clear is that digital technology *has* a very significant impact, and whether that impact is positive or negative will be up to us.

From the positive-computing perspective, the need for preventative positive-computing design is clear. Preventative design means features found to influence empathy *negatively* (by decreasing it or increasing corresponding callous or aggressive behavior) should be identified, and either removed or redesigned. But there is also promise for more active and dedicated approaches that seek to identify and implement features, or even whole systems, for deliberately increasing empathy in the populations who use them.

Technology poses another, more pervasive potential hindrance to our experience of empathy. One of the most significant differences between face-to-face and technology-mediated communication is the poor support for nonverbal communication, which is so critical to empathy. Many of the environmental and physical cues—such as gestures, facial expressions, and tone of voice—that critically inform and shape our understanding of what others are feeling in a face-to-face context are entirely missing from computer-based communication. Mihaly Csikszentmihalyi suggests that the deep implications of this problem can be understood as a lack of plurality, which inhibits our ability to grow as human beings from our interactions with others (see his sidebar in this chapter).

As such, perhaps the greatest prerequisite challenge regarding empathy for positive computing will be picking away at the barriers to holistic communication imposed by technology design as it exists today. Some will do so by moving toward technology-based interaction that more closely resembles the face to face. Others will advance the capability of computer systems to detect nonverbal cues in order to amplify or reexpress those in other ways. Others will seek nontechnical answers by interrogating our sociocultural expectations of what technology can or should do for us. Whatever the path, we all will need a way of evaluating our efforts, and, to that end, we turn to methods of measuring empathy.

Measuring Empathy

A lack of consensus over a precise definition of empathy has complicated a consensus on valid measures for it. Nevertheless, many measures do exist, some tailored to specific demographics or contexts, others more general. Some measure one particular aspect of empathy (just affective or just cognitive aspects), but the most commonly used self-report measure, the Interpersonal Reactivity Index introduced by Mark Davis (1983) in the early 1980s measures for both.

Another of the most commonly used measures is the Empathy Quotient (Baron-Cohen & Wheelwright, 2004) developed by Simon Baron-Cohen, a leading autism researcher at the University of Cambridge (and yes, he is related to Ali G, but do try and stay focused). The Empathy Quotient is a self-report scale with 60 questions that has been evaluated for its psychometric properties in numerous studies (Lawrence, Shaw, Baker,

Baron-Cohen, & David, 2004). It is a multiple choice questionnaire that, though lengthy, could easily be delivered online. It is often used to measure the impact of therapeutic interventions and is frequently combined with other measures.

Karen Gerdes, Elizabeth Segal, and Cynthia Lietz (2010) point out that self-report measures are by far the most commonly used for measuring empathy, yet there is little they can tell us about empathic *accuracy* (Would you know if you were misunderstanding someone's emotions?), and they therefore recommend validating them with triangulation or comparison methods. They also provide a useful review of both the historical evolution of empathy research and the various measurement strategies available from self-reports to observation and neuroimaging.

Strategies and Interventions for Fostering Empathy

There are many examples of programs for "training" or deliberately developing empathy, sometimes as a facilitative communication skill—for example, in curricula for the education of doctors, nurses, social workers, and other professions. Other times empathy development is conducted as a preventative measure, as for the prevention of mental health problems, bullying, or violent crime in high-risk individuals. Therapeutic interventions develop empathy in mental health contexts, and some interventions have been designed to reduce prejudice and discrimination. Clearly, developing empathy promotes, prevents, and mitigates in many positive ways, but is there a direct link to wellbeing?

Certainly the fact that a pathological lack of empathy is disabling and associated with multiple mental illnesses links empathy to wellbeing from an inverse perspective. But even outside of these relatively rare cases, a lack of empathy will inhibit the development of the strong social relationships so often directly linked to wellbeing. One study (Thomas et al., 2007) found that decreases in wellbeing correlated to the decline in empathy that famously occurs with medical students over the duration of their training and residency. Furthermore, if one is to consider that empathy is a pillar of EI, then we can draw on the research that has linked increased EI to increased wellbeing (Gallagher & Vella-Brodrick, 2008).

Interventions with the specific intention of increasing wellbeing through the development of empathy remain to be developed. Until then, for the purposes of positive computing, the value of increased empathy for wellbeing derives, at the very least, from its associated decreases in unhealthy behavior relevant to technology, such as cyberbullying, self-comparison, and envy.

The literature on interventions for empathy development is of mixed scientific sophistication. On the one hand, there are clear operational guidelines, such as Roman Krznaric's "Six Habits of Highly Empathic People,"[1] but although the veracity of these guidelines might be considered self-evident, they lack scientific backing. A subset of empathy interventions has been thoroughly evaluated in peer-reviewed journals, and an even smaller subset has been evaluated in independent randomized control trials. Studies have demonstrated some of the challenges to carrying out independent evaluations of socioemotional learning interventions.

In one study (Owens, Granader, Humphrey, & Baron-Cohen, 2008), researchers compared LEGO therapy with the Social Use of Language Programme and with a no-intervention control group. The programs were evaluated as social skills interventions for school-age children with high-functioning autism and Asperger's syndrome but are similar to those used for general populations, and the results revealed the importance of using multiple measures for these kinds of studies. For example, on an autism-specific social interaction score (the Gilliam Autism Rating Scale), the LEGO therapy group improved more than the other groups. On a maladaptive behavior score, both the LEGO and the Social Use of Language Programme groups fared better than the control group, and for communication and socialization skills there was a nonsignifiant trend.

Mind Reading: The Interactive Guide to Emotions (created by Simon Baron-Cohen) is a computer-based multimedia program designed to help develop empathy (particularly for those with autism spectrum disorders). Video, audio (i.e., voice intonation), and storytelling are used to demonstrate an impressive 412 emotions in 24 categories. As a bonus, user progress is rewarded with the opportunity to manipulate the emotional response of a young Daniel Radcliffe (the actor who played Harry Potter).

Whereas Mind Reading is typically used as a therapeutic intervention, Roots of Empathy is a preventative measure.[2] Roots of Empathy is a highly successful school program for which babies and their mothers visit a classroom on a regular basis throughout the year. The children in the class develop empathy by observing the baby's emotions and behaviors in response to toys, to their mother, and to the class. This program has been shown to have a significant impact on reducing bullying and violence in schools (which suggests that online interventions for empathy could reduce online bullying and aggressive behaviors).

An intervention such as Roots of Empathy is decidedly not technological and, as such, is inspirational in that it shows how very simple and natural human interactions are enough to increase empathy significantly in developing children and to lead to concrete results in the form of decreased antisocial behavior. Just as adding names and subjectivity to the story of mice at the start of this chapter increased reader empathy for the mice, and just as photographs of people in news stories increase our connection with political issues, fostering empathy does not have to be complicated or high tech. It is a natural product of human biology for which simple nudges and experiences can go a long way. We discuss low-tech approaches to developing empathy within digital environments later under "Design Implications."

Of course, there are more difficult cases, such as when ingrained historical prejudices or conflict obstruct empathy development, and in these cases more advanced technologies are already showing they may have a very powerful role to play.

Technologies for the Development of Empathy

Computers themselves demonstrate a total lack of empathy. We get not so much as an apologetic nod when they crash and lose days of hard work. Although designers have become more adept at hiding this lack of empathy through more creatively written error messages and elegant interface interactions, the reality remains that our personal computers currently have no ability to understand what we are feeling or to react appropriately.

Of course, it is this challenge to which many researchers in affective computing have turned. Affective-computing researchers work on building technologies that can recognize and in some cases even express emotions as feedback for their human users. Although we suggest caution with designing machines to solicit or express empathy in human ways (which could become emotionally confusing or potentially undermine our ability to set sensible priorities), there are already some applications of virtual empathy that show very promising results for positive computing (see the work of Timothy Bickmore and his sidebar at the end of this chapter). There is also a clear role for this research to help technologies respond better to users as well as to help us understand human empathy better. Rana El Kaliouby, Rosalind Picard, and Simon Baron-Cohen (2006), for example, have acknowledged the overlap of goals and challenges between autism research and affective-computing research and suggest that closer collaboration will lead to benefits to both areas.

In the computer-supported education context, one study (Cheng, Chiang, Ye, & Cheng, 2010) has given evidence for the efficacy of using collaborative 3D virtual learning environments to enhance empathy. This program, like the Mind Reading program discussed earlier, was developed for people with empathy-related special needs. With the move from desktop to touch-screen computers, doors have burst wide open in the area of special-needs support. For example, the improved usability and increased access made possible by mobile technologies has fueled incredible growth in the development of apps for children with disabilities (the advocacy group Autism Speaks maintains a list of dozens of apps for supporting children with autism at its website autismspeaks. org). Moreover, there are also an increasing number of apps, games, and other technologies for developing factors such as empathy for the general population.

Developing Empathy through Gaming

Common Sense Media is a nonprofit organization dedicated to providing trustworthy media reviews to help parents make informed choices about family media consumption. It rates various forms of media on scales for criteria such as positive role models, violence, scariness, and consumerism. Along with a list of books and movies that promote empathy, they have a list of top games for empathy building. These games target a variety of platforms from handheld devices and console games to mobile and computer-based games. *Herotopia*, for example, is a computer game in which kids combat bullying. Kids at Home is an app that allows kids to learn about diverse cultures and lifestyles by visiting other kids' homes virtually, and *Mission US: Flight to Freedom* is a computer-simulation game in which players experience the life of a slave girl in pre–Civil War America. The diversity and creativity in these early examples is inspiring, and games for empathy are by no means restricted to children's titles.

Games have some magical powers when it comes to supporting empathy. They have the capacity to provide us with "firsthand" (even embodied) experience of scenarios that would otherwise remain totally foreign to us.

As vehicles for role playing, they allow us to "walk a mile in another man's shoes" as close to literally as possible. Combine these affordances with the growing work in games for social good, and a new era of empathy-promoting games becomes visible on the horizon.

For example, browse the selection at GamesforChange.org, and you'll find that making social change requires empathy and that many of these games are designed to promote it. Jonathan Belman and Mary Flanagan (2010) provide a review and insight into designing games that foster empathy, along with a series of design heuristics that we'll come back to in the design section of this chapter.

Role-playing games designed to promote conflict resolution have gotten special attention for their ability to allow players not only to envision but to take first-person action from the perspective of two different sides of a conflict. For example, in the multi-award-winning *PeaceMaker* game,[3] players variably take on the role of Israeli prime minister and Palestinian president, and in both conditions they must choose strategies to make a "two-state solution" viable. The players have the choice to take conciliatory or hostile actions and to pursue collaborative or one-sided initiatives. The game provides players with real-life consequences to political decisions and shows how even small gestures can contribute to peaceful solutions. It incorporates real photos and video, which not only lend the experience authenticity but also engage players emotionally (i.e., empathically).

The makers of *Frontiers*, a 3D online multiplayer game, took the popular *HalfLife* 2D game (a first-person shooter) and modified it into a moving and realistic immersion into the migration paths and borderlands inhabited by political refugees. Players can elect to play the role of an escaping refugee or a member of the border patrol. The game makers, a group of Austrian artists, describe *Frontiers* as both a game and a work of art that "aims to enhance the perception and understanding of the migrants situation above a causal level of catastrophic news."[4]

These are just a few examples of games designed to foster empathy. The number of such games in the name of art, entertainment, and social change is on the rise, and we hope to see a corresponding increase in evaluations of the impacts of these games in the research literature in the near future. Certainly, the literature demonstrating the prosocial benefits of playing prosocial games is relevant here. Tobias Greitemeyer, Silvia Osswald, and Markus Brauer (2010), for example, have given evidence that prosocial games increase interpersonal empathy and decrease schadenfreude.

Further along the immersion scale, the power of highly embodied virtual experience to increase levels of empathy and even altruistic behavior has been indicated by research in virtual reality (see Jeremy Bailenson's sidebar on virtual reality for empathy and prosocial behavior in chapter 11). Because these studies have measured an increase in prosocial action (rather than specifically empathic response), we come back to them in the following chapter on compassion and altruism.

Research Technologies for Understanding Empathy

Beyond promotional or therapeutic interventions, technologies can also be used to help us better *understand* factors of wellbeing such as empathy. Computer vision tools and affective-computing techniques, for example, have been used to explore parent–child interactions and how empathy is developed (Messinger et al., 2014). Although we have not included these types of applications within the scope of positive computing, we believe they represent a closely related field of research in which psychologists and engineers will need to work together.

Technologies for Group Empathy

Although the term *empathy* is most frequently used in the context of one-on-one interaction, it can also be extended to groups of people. Certainly, we can empathize with a group of people when we come to know something of their suffering. Likewise, it seems reasonable to assume that good leaders and, in an organizational context, effective managers will empathize with the joys and struggles of their team. Take, for example, David Caruso and Peter Salovey's (2004) notion of the emotionally intelligent manager. Arguably, managers and leaders need an ability not only to empathize one on one, but also to attend to the pulse of a team's or organization's emotional state. We suspect that technologies have a unique capacity to make this feasible—in other words, to help leaders to better understand the collected emotions of a large *group* of people and thus to be able to react to them more appropriately (e.g., through better management practices or policies). The benefit of such technology is even more clear in the modern workplace in which coworkers are distributed over time and space and meet only on occasion and in which most communication occurs digitally.

Imagine this scenario:

Mr. W. E. Coyote is a director at ACME Inc., an innovative company with 1,200 employees. The fast-growing company has recognized employee wellbeing as important, both for its own sake and as a way to increase productivity and staff retention. Mr. Coyote would like to measure the impact that his new policies have on the subjective wellbeing of his staff, including their engagement and life satisfaction. Fortunately, the company has an active community in its Enterprise Social Network, and the forums in its knowledge-management system are thriving. He can't help but think there must be some way he could use this information—which employees make publically available to the whole organization—to gauge the impact of his policy and management interventions.

Can technology help organizational decision making by supporting EI with information about employees' emotional state? We think it can (Calvo, Pardo, & Peters, 2013), and, indeed, companies such as Kanjoya are already commercializing products that use data from customer forums and enterprise social networks to turn "unstructured data and digital texts into actionable insights,"[5] communicated through various visualizations, in order to help organizations improve leadership decision making and customer experience.

New data algorithms can also be trained to detect people's emotions based not only on written text, but also from speech, writing, facial expression, voice, and physiology. Behavioral data (e.g., time factors) could be integrated to highlight possible anomalies in organizational "circadian rhythms" (i.e., too many employees working late hours, a likely precursor to poor performance and retention).

We have also speculated on the potential for similar tools to be used in the other direction—toward developing leader self-awareness. For example, Daniel Goleman (2000) identifies six leadership styles:

1. **Coercive** leaders demand compliance: *"Do what I tell you."*
2. **Authoritative** leaders mobilize people toward a vision: *"Come with me."*
3. **Affiliative** leaders seek to create harmony and build emotional bonds according to the general principle *"People come first."*
4. **Democratic** leaders forge consensus through participation and inquiry, by asking, for example, *"What do you think?"*
5. **Pace-setting** leaders set high standards by example: *"Do as I do, now."*
6. **Coaching** leaders place the focus on developing people for the future. The message for these leaders can be summarized as *"Try this."*

Leadership styles, whether more authoritarian or more permissive, impact employee performance and even wellbeing. Therefore, helping leaders to recognize these elements in themselves can help them to leverage their strengths, change unhelpful approaches, and adapt to various contexts and circumstances as appropriate.

Design Implications

Technological Communication
As mentioned earlier, one of our greatest design challenges will be devising ways to manage, technically resolve, or compensate for the lack of empathy cues and plurality inherent to current technology-mediated communication (particularly if it continues to comprise an increasing percentage of our interaction with others).

In theory, adding more accurate sensory channels (video, audio, tactile, olfactory, etc.) should begin to reduce the social filter, but current technologies are still far from matching face-to-face interaction, and, perhaps, rather than attempting to force various digital media to approximate face-to-face interaction, we should be looking to medium-specific strategies for better results. The uptake of emoticons is a perfect example of a crowd-based design solution to computer limitations.

In a chat box, I can't indicate to you that I'm sad, excited, or being facetious, so I append representative emotional symbols to compensate. When hoping for some sympathy, the tearing sad face does the trick. We have a friend who insists there should be a typographic style (like italics or bold) for indicating sarcasm. Along these lines, one approach to improving our ability to express emotion and empathize via online interaction may come in the form of more sophisticated approaches to annotation and labeling.

Preventative and Active Design Approaches
Custom technologies dedicated to fostering empathy will make up only a percentage of work in this area, whereas strategies integrated into other kinds of software have the potential to reach wider audiences. Finding ways to prevent loss of empathy or to support it in the context of other online activities could prove beneficial to large numbers of developers across multiple industries. For example, as we begin to separate those aspects of social networks that decrease empathy from those that are neutral or promotional, we can begin to redesign interfaces and interaction for a better user experience. These measures could decrease negative experiences such as cyberbullying but also increase positive experiences via increases in positive emotions and empathic joy.

Likewise, as we begin to separate those aspects of digital games that decrease rather than increase empathy, we can begin to lead a shift in game design that embraces positive outcomes and player wellbeing and moves away from design features with measurably deleterious effects. Imagine a generation of people who grew up on games that developed their empathy for others and helped them learn to care for and collaborate with each other—we'd sign up for that future.

Low-Tech Approaches—the Power of Graphics and Narrative
Some empathy design interventions will remain decidedly low tech. Already, designers frequently add high-impact photographs of people and faces to communicate the humanity behind otherwise faceless stories and issues. By way of example, charity organizations have been moving from an older approach of including imagery of those suffering in extreme ways to including imagery of those being empowered by charitable assistance to improve their lives, regaining joy and autonomy (e.g., a child dying of malnutrition versus a newly thriving community tending to the livestock provided by charitable donation).

Oversensitivity to the feelings of others can be considered a downside or malfunction of the regulation aspects of empathy. A feeling of being overwhelmed or an inability to cope in the face of another's suffering can lead to depression, hopelessness, and lack of action. Allowing for empathic joy in the face of great social challenges adds empowerment to the picture, moving empathic concern toward practical action. Based on appraisal studies, it has been noted that "feelings of compassion should increase when the individual feels capable of coping with the target's suffering. Appraisals of low coping ability, by contrast, should activate distress in the face of another's suffering, which countervails compassion-related tendencies when resources are low," and that "sadness and fear are associated with appraisals of feeling weak, powerless and unable to cope" (Roseman, Spindel, & Jose, 1990 and Hoffman 1981, quoted in Goetz, Keltner, & Simon-Thomas, 2010). This really crosses over to the subject of compassion and altruism, which we move to in the next chapter. Nevertheless, the example of charities' image choice shows the emotional power that simple design decisions can have.

Moving to the workplace, we come across many contexts in which human strangers must interact with each other online to conduct business and without the benefit of face-to-face interaction. In customer-service scenarios, such as call centers, both customers and representatives may find it difficult to empathize with each other, increasing the chances of impatience, customer/job dissatisfaction, or conflict. Simply increasing information channels (such as moving to video chat) would be intrusive in this context, but low-tech solutions, such as showing customer-type images to reps for each call or vice versa—displaying rep images to the customer for online interactions—could prove useful strategies for improving empathy and overall wellbeing.

High-Tech Approaches—Role Playing and Embodiment
In light of early research in digital games and virtual reality, it seems reasonable to imagine that these technologies' ability to support role play and full-body experience will have significant impact on the use of technology for developing empathy. Embodiment, whether as part of virtual-reality environments or gesture-based videogame play, has the unique ability to activate the empathy-promotional benefits of synchronous movement and to allow people to experience, as close to literally as possible, lives, circumstances, and times that would otherwise be impossible for them to experience.

In one study, Peter Yellowlees and James Cook (2006) created a virtual psychiatry clinic in the virtual world *Second Life* to promote empathy for people with serious mental illness. By touring the clinic, participants could experience firsthand the auditory and visual hallucinations associated with psychosis. The majority of more than 500 visitors to the clinic who voluntarily responded to the researchers' survey said it helped them better understand this experience.

Heuristics for the Design of Empathy Games
Belman and Flanagan (2010) speculate on a state of game play they call "empathetic play" in which "players intentionally try to infer the thoughts and feelings of people or groups represented in the game (cognitive empathy), and/or they prepare themselves for an emotional response, for example by looking for similarities between themselves and characters in the game (emotional empathy)." They further speculate that games for empathy must deliberately support the player in entering a state of empathetic play and that it should not be assumed that the content is enough to elicit empathy as intended. This recommendation is formulated as one of their four principles for the design of games to foster empathy. They draw these principles from their experience working with designers of "games for good."

The principles, although pending evaluation at the time of this writing, are founded in current practice and related empathy research:

1. **Induce empathy from the start.** Belman and Flanagan (2010) state that "players are likely to empathize only when they make an intentional effort to do so as the game begins. The game may explicitly ask players to empathize, or it may more subtly encourage them to take on a focused empathetic posture. However, without some kind of effective empathy induction at the outset, most people will play 'unempathetically.'"
2. **Recommend actions.** Belman and Flanagan suggest designers give players specific suggestions as to actions they can take to address issues represented in the game. They speculate that the importance of empowering players to take action may help prevent negative consequences associated with empathic suffering. This also relates back to the connection between low coping ability and empathic distress. Based on these findings, Jennifer Goetz, Dacher Keltner, and Emiliana Simon-Thomas (2010) speculate that "variables that enhance a sense of coping should make one more likely to feel compassion than distress."
3. **Design for cognitive and/or affective empathy as appropriate.** If desired outcomes don't require significant changes in player beliefs, then a "short burst of emotional empathy" can work well. However, when deeper changes in thinking are a requirement, then "the game should integrate both cognitive and emotional empathy" (Belman & Flanagan 2010).
4. **Emphasize similarity but with delicacy.** Belman and Flanagan suggest that designers "emphasize points of similarity between the player and people or groups with whom she is supposed to empathize, but beware of provoking defensive avoidance."

We encourage you to see the full paper for a full explanation of these principles along with some exemplary cases.

Technologies for empathy are still in their infancy, but it is an impressively enthusiastic infancy. Apps as a format have been embraced by those seeking to support empathy development in children, and the new wave of designers interested in creating games for social change has fueled a slew of sophisticated vicarious experiences created to deepen our understanding of, concern for, and desire to assist those in dire circumstances. Moreover, as discussed, great art has always played a role in developing empathy and compassion. Now imagine the impact of an experience that converges the literary prowess of Victor Hugo and the artistic genius of Van Gogh with the interaction and embodiment made possible by the modern digital games medium. Should such a transformative convergence take form, expect to see us first in line to play.

Expert Perspectives—Technology for Empathy and Plurality

Compassionate Computer Agents

Timothy W. Bickmore, Northeastern University

Health represents one of the most primal human needs, and technologies designed to positively impact health thus compose an important aspect of positive computing. Over the past decade, many researchers have investigated the development of automated caregivers that promote physical and mental health in people who interact with these systems. Many of these researchers believe that the most promising design methodology is to emulate the best role models we have—expert, compassionate, human caregivers—with as much fidelity as possible. The resulting systems are conversational, communicating with people in some form of natural-language dialogue, often accompanied by a virtual or physical representation of a human caregiver's persona.

Media used in these systems span voice only (interactive voice response), animated agents (also known as "embodied conversational agents"), and social robots. The anthropomorphic qualities of these interfaces not only serve to make them approachable by a wide range of users but are essential for conveying many of the subtle cues used by human caregivers, such as empathy and compassion. These systems are necessarily limited in scope, but they promote some narrow dimension of health and wellbeing among their users, typically over the course of many conversations that can last weeks or months.

(continued)

> Most of the automated caregivers developed to date address physiological and mental health needs, including interventions for medication adherence, exercise and diet promotion, cancer prevention, prenatal care, and depression counseling. Most of these systems have demonstrated significant improvements in health relative to nonintervention or standard-of-care control groups, and many have been shown to be equal to interventions performed by human caregivers.
>
> An increasing number of automated caregivers are being developed whose primary goal is to provide companionship, typically for older adult users who live alone. Although these agents may provide health or wellness coaching or cognitive stimulation activities, they also engage in a wide variety of social behaviors whose goal is to provide a feeling of social support among their users, which has been shown to be an important factor in decreasing mortality among isolated elderly. One study demonstrated that a companion agent provided to isolated older adults significantly decreased loneliness after a week and that the more users talked to the agent, the less lonely they were.
>
> In addition to general anthropomorphic qualities, these agents use specific human behaviors to improve their effectiveness and their users' sense of wellbeing. Social bonding between a caregiver and client (referred to as "therapeutic alliance") is crucial in maximizing outcomes for all kinds of health interventions. Agents that use behavior designed to improve social bonding and therapeutic alliance have been referred to as "relational agents," and several studies have shown relational behaviors can lead to significant increases in engagement with these agents. One of the most important agent behaviors for social bonding is the display of empathy, comforting, and compassion for the user when they are in a distressed state. Even though these behaviors are simulated and thus do not exhibit "true empathy" (by many definitions), a wide range of agent behaviors—including facial displays of concern, empathic language, and even physical touch—have been shown to significantly reduce user frustration, increase positive affect, and decrease loneliness over time—impacts that are arguably among the most important objectives of positive computing at the individual level.

(continued)

Computers and the Condition of Human Plurality

Mihaly Csikszentmihalyi, Claremont Graduate University

It's been half a century since Marshall MacLuhan wrote his treatise about the media, which was often wrong-headed but contained some truths as well. As is usually the case, the really important contributions of his thinking seem to have been forgotten. Or at least they have not been applied very much to the new media that have arisen since his writing.

If one just takes the most memorable line from MacLuhan's work, *"the medium is the message,"* and applies it to computers and to the Internet, some interesting issues are foregrounded.

Let us just think about how computer-mediated communication differs from interpersonal face-to-face communication. The latter is the expression of what the social philosopher Hannah Arendt has called the fundamental fact of the human condition: *plurality*. What makes us human is that we can communicate our individual uniqueness to each other, and in so doing we can grow beyond the boundaries set by our genes and early environment. The only reason we can say, with Walt Whitman, that we contain multitudes, is that we are open to information coming from people who are so much like we are, yet also so different.

(continued)

> This information, which makes us who we are, is not contained just in the meaning of words. For us to take the information seriously, we must feel that the sender is to be trusted. And trust depends on many things: we now know it depends in part on how the person in front of us smells as well as on looks, expressions, tone of voice, and so on. After all, we are not going to take seriously messages sent by an unknown source or by someone who is completely different from us.
>
> What is essential to the condition of plurality is this fine balance between similarity and dissimilarity among the partners of the exchange. If there is too much similarity, learning something new is unlikely; if there is too much dissimilarity, the information is unlikely to be relevant. For growth to occur, the information that contradicts our beliefs and prejudices must come from trusted sources.
>
> What is missing from the so-called social media is the condition of plurality. Social media are *social*, but rarely *plural*. In other words, in order to maintain contact on the web, the parties usually try to emphasize their similarity. But because the communication relies only on the written word and not on the looks, smells, and other dimensions of physicality that facilitate openness and trust, each party essentially remains himself or herself, without having the growth-producing experience that an encounter among humans has, at least when the conditions for it are right.
>
> Of course, we can and do learn a great deal from computer-mediated information—especially when the source has been established as trustworthy and when the information we seek is factual and objective. But some of the most important knowledge we seek is existential, or subjective and dialogical—that is, knowledge that helps us become fully grown persons. The computer is still not a very good medium for gaining that kind of information. Will it ever be? I am not a betting man, so I will stop right here.

Notes

1. For these habits, see greatergood.berkeley.edu/article/item/six_habits_of_highly_empathic_people1 or Krznaric's video at goo.gl/7Axbg.

2. For Roots of Empathy, see the website at rootsofempathy.org.

3. *PeaceMaker* can be found at peacemakergame.com.

4. From the game makers' website at goldextra.com.

5. For Kanjoya, see kanjoya.com/crane.

References

Anderson, C. A., Shibuya, A., Ihori, N., Swing, E. L., Bushman, B. J., Sakamoto, A., et al. (2010). Violent video game effects on aggression, empathy, and prosocial behavior in eastern and western countries: A meta-analytic review. *Psychological Bulletin, 136*(2), 151–173.

Barker, R. L. (2008). *The social work dictionary*. Washington, DC: NASW Press.

Baron-Cohen, S., & Wheelwright, S. (2004). The empathy quotient: An investigation of adults with Asperger syndrome or high functioning autism, and normal sex differences. *Journal of Autism and Developmental Disorders, 34*(2), 163–175.

Belman, J., & Flanagan, M. (2010). Designing games to foster empathy. *International Journal of Cognitive Technology 15*(1), 5–15.

Buckley, K., & Anderson, C. (2006). A theoretical model of the effects and consequences of playing video games. In P. Vorderer & J. Bryan (Eds.), *Playing video games: Motives, responses, and consequences* (pp. 363–378). New York: Routledge.

Calvo, R., Pardo, A., & Peters, D. (2013). Supporting emotional intelligence with automatic sensing of the organization's affective state. Presented at: Emotional intelligence in the workplace. May 29–31, London.

Caruso, D. R., & Salovey, P. (2004). *The emotionally intelligent manager: How to develop and use the four key emotional skills of leadership*. San Francisco: Jossey-Bass.

Cheng, Y., Chiang, H.-C., Ye, J., & Cheng, L. (2010). Enhancing empathy instruction using a collaborative virtual learning environment for children with autistic spectrum conditions. *Computers & Education, 55*(4), 1449–1458.

Davis, M. H. (1983). Measuring individual differences in empathy: Evidence for a multidimensional approach. *Journal of Personality and Social Psychology, 44*(1), 113–126.

Decety, J., & Moriguchi, Y. (2007). The empathic brain and its dysfunction in psychiatric populations: Implications for intervention across different clinical conditions. *BioPsychoSocial Medicine, 1*, 22.

De Vignemont, F., & Singer, T. (2006). The empathic brain: How, when, and why? *Trends in Cognitive Sciences, 10*(10), 435–441.

El Kaliouby, R., Picard, R., & Baron-Cohen, S. (2006). Affective computing and autism. *Annals of the New York Academy of Sciences, 1093*, 228–248.

Farrant, B. M., Devine, T. A. J., Maybery, M. T., & Fletcher, J. (2012). Empathy, perspective taking, and prosocial behaviour: The importance of parenting practices. *Infant and Child Development, 21*(2), 175–188.

Gallagher, E. N., & Vella-Brodrick, D. A. (2008). Social support and emotional intelligence as predictors of subjective well-being. *Personality and Individual Differences, 44*(7), 1551–1561.

Gentile, D. A., Anderson, C. A., Yukawa, S., Ihori, N., Saleem, M., Ming, L. K., … Bushman, B. J. (2009). The effects of prosocial video games on prosocial behaviors: International evidence from correlational, longitudinal, and experimental studies. *Personality and Social Psychology Bulletin, 35*(6), 752–763.

Gerdes, K., Segal, E., & Lietz, C. (2010). Conceptualising and measuring empathy. *British Journal of Social Work, 40*(7), 2326–2343.

Goetz, J. L., Keltner, D., & Simon-Thomas, E. (2010). Compassion: An evolutionary analysis and empirical review. *Psychological Bulletin, 136*(3), 351–374.

Goleman, D. (2000). Leadership that gets results. *Harvard Business Review, 78*(2), 16–28.

Greitemeyer, T., Osswald, S., & Brauer, M. (2010). Playing prosocial video games increases empathy and decreases schadenfreude. *Emotion, 10*(6), 796–802.

Hoffman, M. L. (1981). Is altruism part of human nature? *Journal of Personality and Social Psychology, 40*, 121–137.

Konrath, S. H., O'Brien, E. H., & Hsing, C. (2011). Changes in dispositional empathy in American college students over time: A meta-analysis. *Personality and Social Psychology Review, 15*(2), 180–198.

Langford, D. J. (2010). *Social modulation and communication of pain in the laboratory mouse*. Montreal: McGill University.

Langford, D. J., Crager, S. E., Shehzad, Z., Smith, S. B., Sotocinal, S. G., Levenstadt, J. S., … Mogil, J. S. (2006). Social modulation of pain as evidence for empathy in mice. *Science, 312*(5782), 1967–1970.

Lawrence, E. J., Shaw, P., Baker, D., Baron-Cohen, S., & David, A. S. (2004). Measuring empathy: Reliability and validity of the Empathy Quotient. *Psychological Medicine, 34*(5), 911–919.

Messinger, D., Duvivier, L. L., Warren, Z., Mahoor, M., Baker, J., Warlaumont, A., ... Ruvolo, P. (2014). Affective computing, emotional development, and autism. In R. A. Calvo, S. K. D'Mello, J. Gratch, & A. Kappas (Eds.), *Oxford handbook of affective computing*. New York: Oxford University Press.

Milgram, S. (1963). Behavioral study of obedience. *Journal of Abnormal and Social Psychology, 67*(4), 371–378.

Owens, G., Granader, Y., Humphrey, A., & Baron-Cohen, S. (2008). LEGO® therapy and the Social Use of Language Programme: An evaluation of two social skills interventions for children with high functioning autism and Asperger syndrome. *Journal of Autism and Developmental Disorders, 38*(10), 1944–1957.

Roseman, I. J., Spindel, M. S., & Jose, P. E. (1990). Appraisals of emotion-eliciting events: Testing a theory of discrete emotions. *Journal of Personality and Social Psychology, 59*, 899–915.

Singer, T. (2006). The neuronal basis and ontogeny of empathy and mind reading: Review of literature and implications for future research. *Neuroscience and Biobehavioral Reviews, 30*(6), 855–863.

Thera, N. (1999). *The four sublime states*. Sri Lanka: Buddhist Publication Society.

Thomas, M. R., Dyrbye, L. N., Huntington, J. L., Lawson, K. L., Novotny, P. J., Sloan, J. A., ... Shanafelt, T. D. (2007). How do distress and well-being relate to medical student empathy? A multicenter study. *Journal of General Internal Medicine, 22*(2), 177–183.

Yellowlees, P. M., & Cook, J. N. (2006). Education about hallucinations using an Internet virtual reality system: A qualitative survey. *Academic Psychiatry, 30*(6), 534–539.

11 Compassion and Altruism

The monster known as Bizarro [was] opposite of Superman in every way, with no compassion, no remorse and no mercy.
—*Superman #23.1: Bizarro* (DC Comics)

To be a superhero is to have compassion, whereas to be a supervillain is to be entirely without it. It's not enough to experience empathy if you're a superhero because heroes need to take action as well. We daydream about being superheroes, not only because it would be great to fly and wear skin-tight unitards, but also because superheroes are fantastically empowered to make things right.

Compassion comes packaged with that all-important desire to act and make change, but less well known is the fact that compassion is also an antidote to the pain and distress that comes with empathy. In other words, compassion can be seen as a form of resilience.

But you were probably expecting a source with slightly more scientific integrity than DC Comics. In terms of the academic literature, compassion, like empathy, suffers from a long-standing lack of consensus with regard to a precise definition. In this chapter, we favor those conceptualizations that view compassion as a distinct emotion based on its unique behavioral and physiological imprint. For example, based on a cross-disciplinary review of the research, Jennifer Goetz, Dacher Keltner, and Emiliana Simon-Thomas (2010) define compassion as "the feeling that arises in witnessing another's suffering and that motivates a subsequent desire to help."

The key words in this definition are *emotion*, *suffering*, and *desire to help*. According to the literature, compassion is considered both an emotional state and a trait (and sometimes a motivation); it is distinguished from love in that it arises as a result of witnessing suffering; it is distinctive from empathy in that the feelings aroused do not necessarily *mirror* those witnessed in another; and it elicits approach behavior and a caring desire to help (which don't necessarily follow empathy).

Although compassion may arise from empathic concern, it can be distinguished by the accompanying desire to take action. This action-oriented aspect of compassion leads us seamlessly into a discussion of altruism. If compassion describes a desire to act, altruism is the action. Altruism is generally described as a type of *behavior* rather than an emotional state. According to much of the literature on altruism, an altruistic act is one that confers benefit on someone else at a cost to oneself (Fehr & Fischbacher, 2003).

As you may suspect, there will be much overlap in the technology examples and strategies we gave in the previous chapter and those found in this one simply because, although empathy and compassion are different, they do, of course, frequently come together. For instance, although the focus of a particular game may be to encourage perspective taking and vicarious feeling (aspects of empathy), this will probably be in aid of inspiring the player to take compassionate action. Nevertheless, it's important for technology designers to distinguish the two because empathy can lead to either wellbeing or distress, and compassion may be the difference.

Any discussion of compassion should include compassion turned toward the self, or self-compassion, which Buddhist psychology implicates as necessary for compassion toward others and which scientific literature has identified as a significant predictor of wellbeing. So, you can also expect some links to chapter 8, herein.

Finally, the mere act of bearing witness to compassion and altruism has its own benefits for wellbeing, so we look at the unique characteristics of inspiration or "elevation" and how technology can be used to promote it.

To our great benefit, the past decade has seen a new wave of psychologists, neuroscientists, and even technologists begin to take on compassion and altruism like never before. There's even a business case for it. Facebook, for example, has hosted several "Compassion Research Days" with the goal of using the science of emotions and relationships to develop features that reduce conflict and increase understanding among people. After all, when people are antisocial or cruel on Facebook, it's bad for business, and if design can make a difference, everyone wins.

This new energy around the science of compassion is likely fueled by the growing interest in wellbeing research in general, but also by movements for social change and on the emerging research showing how compassion could make our world a far better place to live in for everyone. What better reason than to dedicate the following pages to the work of these researchers and to the designers who have followed their lead and begun to explore the capacity for new technologies to play a part in this work?

Research on Compassion and Altruism

When love meets suffering, compassion arises.
—Jack Kornfield, *The Wise Heart*

Compassion, Empathy, Love, Sadness—What's the Difference?

It can be difficult to disentangle compassion from other similar and interrelated emotions such as love, empathy, pity, and sadness. However, the evidence for its uniqueness is summarized elegantly in the review by Goetz, Keltner, and Simon-Thomas (2010). Their synthesis draws on findings in psychology, evolutionary theory, and neuroscience to highlight a number of important ways in which compassion is distinctive, including the following characteristics:

• Whereas empathy is a vicarious experience or mirroring of feeling (positive or negative), compassion is a reaction to another's suffering that does not necessarily entail feeling the same emotion. In other words, if you're angry and I empathize with you, I feel angry, too. On the other hand, if I see through your anger to the hurt behind it, and out of concern am moved to act on your behalf, I have experienced compassion. (More recent research has shown that compassion even activates affiliative positive emotions, which possibly accounts for its ability to promote caring resilience in the face of suffering. We discuss this later in the chapter.)
• Compassion is distinguished from pity in that pity is associated with an appraisal of dominance (feeling in a higher position than the person pitied) and different display behaviors.
• Compassion is characterized by other-centeredness, approach behavior, and action (empathy, particularly in the case of empathic distress, can lead to avoidance).
• Compassion has recognizable facial expressions and display behaviors.
• Compassion has physiological correlates such as reduced heart rate and reduced skin conductance that separate it from empathic distress.
• Although the neuroscientific line of inquiry into the neural correlates of compassion is still nascent, early evidence shows compassion is neurologically distinguishable from similar emotions such as love and sadness.

- Compassion is influenced by a judgment of fairness or justice (we are less likely to be moved by compassion if we view someone's suffering as deserved).

Grit Hein and Tania Singer (2008) point to another stirring difference between empathy and compassion. Pointing to theory of mind (one's ability to know what another is thinking) and how this cognitive aspect can be separate from the *shared-feeling* aspect of empathy (a fact we discussed in relation to psychopathy in chapter 10), the authors explain: "Empathy can have a dark side, for example when it is used to find the weakest spot of a person to make her or him suffer, which is far from showing compassion with the other. It is suggested that empathy has to be transformed into sympathy or empathic concern [we would suggest compassion] in order to elicit prosocial motivation." Sympathy is often used in a way that is roughly synonymous with compassion, especially before the twentieth century. For example, Adam Smith, Charles Darwin, and David Hume used the term *sympathy* (Hein & Singer, 2008; Wispé, 1986).

We propose that the combination of concepts we have just described might be visualized as something like what's shown in figure 11.1.

Paul Gilbert (2013), originator of compassion-focused therapy, describes compassion as a motivation rather than an emotion. Physiologically, compassion preps the body for approach and caregiving. In contrast, empathic distress may urge us to avoid, escape, or close our eyes to curb the pain caused by our response to another's suffering. As discussed in the previous chapter, one's ability to help someone in distress seems to influence whether one takes compassionate action or remains stuck in empathic

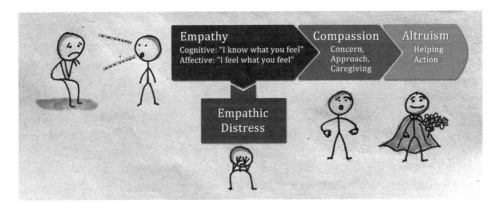

Figure 11.1

distress. Therefore, a question for compassion interventions becomes, Can we foster feelings of agency and empowerment in order to heighten resilience to empathic distress and increase the likelihood of a compassionate response? Research has recently shown that the practice of loving-kindness meditation can reduce the pain of empathy and promote resilient compassion (Klimecki, Leiberg, Lamm, & Singer, 2013).

The ability for compassion to work as a healthier, more effective alternative to empathic distress is also evident in the brain. Olga Klimecki and colleagues have found that while empathy training activates distress and associated neural networks, compassion training elicits activity in brain regions associated with positive affect and affiliation. They concluded that "findings suggest that the deliberate cultivation of compassion offers a new coping strategy that fosters positive affect even when confronted with the distress of others" (Klimecki, Leiberg, Ricard, & Singer, 2013). The implications for preventing burnout, supporting resilience, and promoting wellbeing are significant.

Compassion as Mediated by Fairness

Meg studied hard all night but failed her exam. Mog didn't bother studying, chose to play poker all night instead, and also failed her exam. Who are you inspired to help? You're probably even a bit irritated by Mog. Unsurprisingly, research has found that whether we feel compassion or anger for someone who is suffering is determined by our appraisal of their *deservedness*, or how responsible for their condition we feel they are. As Goetz, Keltner, and Simon-Thomas (2010) put it, "Studies indicate that appraisals of low controllability and responsibility on the part of the target are critical to the elicitation of compassion and not anger."

The influence of blame has implications for those seeking to leverage technology for social change. Prejudice, stigma, and perceived difference often lead people to attribute blame to those who are suffering. For example, various studies have shown that people are likely to blame (and therefore less likely to feel compassion for) people who are suffering as a result of homelessness, obesity, or drug abuse. Tackling the perceptions that sit at the root of blame will be prerequisite to fostering compassion around similar issues.

Fairness also makes an appearance in the literature on altruism. According to research, there is a cross-cultural human value for fairness (likely owing to the evolutionary benefits of a cooperative society), and this desire for fairness is manifest in what researchers call "altruistic punishment." Laboratory experiments that engage people in gamelike interactions in which real money is at stake have shown that people are frequently willing to take action that punishes someone else for being unfair at a cost to themselves (Fehr & Fischbacher, 2003).

This example of "altruistic punishment" is not the kind of warm-hearted, generous, prosocial act one generally associates with the word altruism. Of course, in part that's because altruism is dealt with in research as a behavior rather than as a virtue. At the same time, because punishing the selfish helps enforce prosocial norms (fairness, justice), altruistic punishment can be viewed as benefiting a social good.

Despite the very familiar effects of fairness on both compassion and altruism, there remains a conspicuously missing piece to the puzzle. If an appraisal of fairness is essential to both our experience of compassion and our altruistic behavior, how does one explain those people (often considered the wisest in our societies) who manage to experience compassion and act altruistically despite incredible unfairness and in parallel with appraisals of dramatic injustice?

Compassion in the Face of Injustice
Clearly, for most of us most of the time, fairness has a strong and predictable influence on the likelihood of compassion arising within us. Yet there are people who work compassionately with drug abusers and criminals, and there are widely admired cultural heroes who maintain compassion resolutely in the face of extreme oppression. Buddhist psychology provides a lucid explanation for this phenomenon by explaining that nonjudgmental compassion is not an exception to the rule but in fact describes the very nature of compassion itself, visible when our response is unmediated and uninhibited by the effects of other emotions such as anger, hatred, and fear.

Buddhist psychology defines compassion as an empathic concern that is, by nature, *unaffected by negative behavior* in another (His Holiness the 14th Dalai Lama, n.d.). It is other emotions (i.e., anger, hatred, fear, desire, sadness etc.) that respond to stigma, righteousness, and notions of deservedness, and these emotions can inhibit compassion. The Dalai Lama explains,

We must be clear about what we mean by compassion. Many forms of compassionate feeling are mixed with desire and attachment. For instance, the love parents feel of their child is often strongly associated with their own emotional needs. ... True compassion is not just an emotional response but a firm commitment founded on reason. Therefore, a truly compassionate attitude towards others does not change even if they behave negatively.

... When you recognize that all beings are equal in both their desire for happiness and their right to obtain it, you automatically feel empathy and closeness for them.

His definition speaks of compassion as involving an attitude, a motivation, a cognition, and an emotion. Specifically, it includes "commitment" (motivation), which involves "reasoning" (i.e., perspective taking, appraisal of one's similarity to others), and empathy. Further, he emphasizes that although "developing this kind of compassion is not at all easy!" it is decidedly possible. The deliberate cultivation of compassion is now being played out in Western psychology, and Buddhist compassion practices are proving effective in various contexts from promoting resilience among care workers to treating mental health problems such as anxiety and depression (we discuss some of this work in the section on interventions in this chapter).

One poignant example of compassion as essentially nonjudgmental is the Buddhist practice of cultivating it for one's enemies. Jack Kornfield (2011) describes an encounter with a group of Tibetan nuns who when they were teenagers had survived years of imprisonment, torture, and depravation as part of Chinese government oppression. Asked if they were ever afraid during these years, they responded "Yes, we were terribly afraid. And what we were afraid of was that we would end up hating our guards—that we would lose our compassion. That is the thing we most feared."

Few of us have experience with such a highly developed level of compassion, so the notion that it is possible *not* to feel hatred for torturers (let alone one's own) is hard to fathom. Moreover, it's worth acknowledging that emotions such as anger and fear are useful survival instincts and that their existence alongside empathy has probably helped us protect ourselves. However, relying on these emotions to guide our actions does confer a cost. We know emotions such as anger and fear are hard on the mind and body, and when they carry on unregulated, they become hatred or chronic anxiety. Furthermore, they motivate quick but not necessarily reasoned or wise action.

The monks who cultivate compassion for their enemies do appraise their situations as unjust, and they actively struggle to end the injustices (Buddhists from Tibet, Burma, and previously Vietnam are examples), but at the same time they work to reduce the development of self-destructive emotions such as hatred and terror—emotions that can hinder both individual resilience and wise action.

From this perspective, the nuns' admission that losing their compassion was their greatest fear seems incredibly reasoned. Their compassion was neither a denial of circumstances nor blind polyannaism, but rather a key to their resilience. Losing it would have meant losing their greatest asset in surviving such trauma. Matthieu Ricard (2007), among others who have worked intimately with Tibetan refugees, observes that this population shows fewer symptoms of post-traumatic stress as a result of their spiritual practice (see also Elsass, Carlsson, & Husum, 2010).

Neuroplastic changes have been demonstrated in neuroimaging studies that differentiate the brains of expert and novice practitioners of a number of meditative practices, including compassion cultivation (Lutz, Brefczynski-Lewis, Johnstone, & Davidson, 2008).

The ability to cultivate compassion in a way that is disentangled from associated self-destructive emotions is utterly necessary if compassion is to support emotional resilience *even* in the face of difficulty and injustice. The Buddhist approach presents us with the possibility that we can develop our capacity for compassion to a point where the primal responses of anger and fear have less power and where compassion can serve as a healthier regulator of action—a cognitive-emotional experience that both builds resilience and increases our wellbeing. The example of the nuns seems to describe compassion training at Olympic proportions, but their existence as well as research studies on compassion practice give evidence that cultivating compassion is possible and that each step on the path increases wellbeing.

Paul Gilbert (2005) for example has developed methods for people to increase their compassion in order to improve their mental health. Gilbert and Sue Procter (2006) define compassionate abilities as including "nonjudgment related to the ability to be non-critical of the other's situation or behaviors." They add that all the aspects of compassion "require the emotional tone of warmth." This "nonjudgment" and "warmth" turn out to be critical to the wellbeing benefits of self-compassion and elements previously missing from many other psychotherapeutic approaches.

Altruism—Why Bother?

One question that has intrigued scientists for centuries is why humans would be moved to help others without benefit to themselves and even at great cost. Although colloquially we think of altruism as being defined by benevolence and a refreshing lack of ulterior motive, many researchers have been little satisfied with such an explanation and have made a case for ulterior motives as diverse as expected reciprocity, reputation building, peer pressure, fear of punishment, and increasing one's chances of mating (a number of studies show that indicators of compassion and altruism increase one's perceived attractiveness). Social exchange theory (Homans, 1958) suggests that all behavior (altruism included) is regulated by cost–benefit analysis (and made on the basis that benefit will outweigh cost).

Theories of altruism based on egoistic motivators are an important contribution to the research. After all, we are indeed influenced by social pressure, and compassionate mates make better parents, so it's sensible they would attract. But we can't help but feel that if we left you with a sense that fostering altruism was just about helping your users get their mojo on, it would quickly lose its appeal as a potential focus for positive computing. Fortunately, there are also research-based conceptualizations that indicate altruism just might be motivated by empathy and compassion and that describe it as an other-centered phenomenon.

In another of Charles Darwin's way-ahead-of-his-time moments, he described compassion (he used the term *sympathy*) as stronger than any other of our social instincts and as utterly logical from an evolutionary standpoint because "sympathy will have been increased through natural selection; for those communities, which included the greatest number of the most sympathetic members, would flourish best, and rear the greatest number of offspring" (from *The Descent of Man, and Selection in Relation to Sex*, quoted in Goetz et al., 2010).

To put it simply, compassion and altruism are evolutionarily beneficial because they are prosocial: they allow for collaboration within one's group, but also for cooperation with those outside of it. Compassion also allows for the caring of vulnerable offspring, which not only ensures greater group survival but would have been necessary for increasingly intelligent humans to survive the corresponding period of helplessness required by a larger brain (a newborn giraffe can stand up and run within an hour of being born, but humans are utterly dependent for years as their complicated brains mature). The Dalai Lama (n.d.) similarly articulates the evolutionary imperative of compassion: "The need for love lies at the very foundation of human existence. It results from the profound interdependence we all share with one another. However capable and skillful an individual may be, left alone, he or she will not survive. However vigorous and independent one may feel during the most prosperous periods of life, when one is sick or very young or very old, one must depend on the support of others."

Thomas Aquinas, David Hume, and Adam Smith are among those who have pointed to empathy and compassion as a trigger for altruism. More recently, in the 1980s, C. Daniel Batson began to explore ways of testing motivations behind altruism experimentally in order to disentangle the various possible motivations (self-focused or other-focused). As a result of his findings (Batson, 1991, 2002), Batson proposed the "empathy–altruism hypothesis," which states that the likely cause of altruistic motivation is empathy.

That's not to say that all altruistic behavior is always triggered exclusively by empathy, but Batson's work is critical in making the point that nonegoistic altruism does in fact exist. Stanford's Center for Compassion and Altruism Research and Education and Berkeley's Greater Good Science Center are sources of a wealth of relevant research in this area.

One paradoxical and fascinating element of altruism is that when it does stem from an other-centered compassionate response, then it seems to confer significant benefits to the self—namely, measurable increases to wellbeing.

Compassion and Wellbeing

One interesting, even paradoxical, thing about compassion is that despite the fact that it involves witnessing *suffering* (and often leads to action at a *cost* to oneself), it has been shown in many different studies to confer benefits of wellbeing, including increased social connectedness and stress-reduction.

Compassion reduces stress (Cosley, McCoy, Saslow, & Epel, 2010) in a way that doesn't require the kind of escape from events that other forms of stress release do. Whereas empathic distress can lead to elevated heart rate and skin conductance associated with stress (fight or flight responses), compassion responses trigger heart-rate *deceleration* and lowered skin conductance consistent with the brain prepping for caregiving and other-focused attention (Goetz et al., 2010). In other words, whereas stress prevents us from being able to help others in need (because it triggers fight-or-flight states rather than affiliative states), and not being able to help others in need causes us stress (as in empathic distress), compassion may be an antidote. The stress-buffering effect of compassion versus empathy-related distress is manifest in the phenomena of "compassion satisfaction" versus "burnout" among aid workers (Conrad & Kellar-Guenther, 2006; Thomas, 2013).

If that weren't enough, compassion has also been described as an instrument against fear, anger, envy, and vengeance (Goleman, 2003), and studies have linked compassionate lifestyles to greater longevity (Brown, Nesse, Vinokur, & Smith, 2003; Okun, Yeung, & Brown, 2013). Be nice, live longer. What's amazing is that just the act of volunteering is not enough; you have to be doing it for the right reasons. Morris Okun, Ellen WanHeung Yeung, and Stephanie Brown (2013) found that older adults volunteering for *self-oriented reasons* had mortality rates similar to nonvolunteers, whereas those who volunteered for *other-centered* reasons experienced lower mortality. This suggests it is compassion itself that contributes to our wellbeing, perhaps owing to its very beneficial physiological effects.

Researchers have hypothesized that compassion's effect on wellbeing might also be explained by the way it supports social connectedness and/or by the way it helps us to broaden our perspective beyond ourselves (both depression and anxiety are linked to self-focus). (For an accessible review of the literature behind compassion and its influence on wellbeing, see Seppala, 2013.)

If the design of technologies can encourage the development of compassionate attitudes and can elicit compassionate states in the face of social problems, we will not only be helping to address those problems, we'll also be improving the wellbeing of those experiencing compassion.

Self-Compassion and Wellbeing

Although we have consistently described compassion as "other-centered," this attitude can also (more paradoxes) be directed toward the self and with remarkably positive consequences. As mentioned previously, self-compassion "entails being kind and understanding toward oneself in instances of pain or failure rather than being harshly self-critical; perceiving one's experiences as part of the larger human experience rather than seeing them as isolating; and holding painful thoughts and feelings in mindful awareness rather than over-identifying with them" (Neff, 2011). When people can react toward themselves in a caring, nonjudgmental way, as they would to a close friend or child, then wellbeing benefits manifest.

The work of Paul Gilbert has validated the effectiveness of cultivating compassion for the self as a method for treating people who suffer from extreme self-criticism and shame. More so than self-esteem, compassion is linked to affiliative physiological responses and warm emotional tone, which, according to Gilbert (2010), are critical in these cases. According to his work compassion generates essential elements of physiologically felt nonjudgmental nurturing and love that are necessary to wellbeing.

Perhaps cultivating these nurturing qualities toward oneself increases their availability in response to others, as both the trait and practice of self-compassion has been correlated with increased compassion for others (Reyes, 2011). The same study shows correlations with increased self-care capacity, relatedness, autonomy, and sense of self. It has also been associated with increased psychological functioning and reduced symptoms of anxiety, depression, and rumination (Neff, Kirkpatrick & Rude, 2007).

Other studies show greater self-compassion predicts lower incidents of automatic thoughts, interpersonal cognitive distortions, and, as mentioned previously, Internet addiction. Could methods for cultivating the components of self-compassion (or reducing their polar opposites) be used within the context of games and other technologies as a measure for decreasing addiction?

Altruism and Wellbeing—in Giving of Ourselves, We Receive

Wellbeing benefits extend to altruism as well. Groundbreaking work has shown that, despite what our inner skeptics might assume, human beings (from as young as infancy) derive greater happiness from giving something away (or even from watching things being given to others) than they do from receiving those things themselves (Aknin, Hamlin, & Dunn, 2012; Dunn, Aknin, & Norton, 2008). From another angle, studies have found that across cultures, the amount of money people spend on other people correlates strongly to their personal wellbeing, regardless of their income (Aknin et al., 2013). According to Elizabeth Dunn and Michael Norton (2013), "across the 136 countries studied, donating to charity had a similar relationship to happiness as doubling household income."

The idea that giving produces positive emotions that exceed receiving could have many positive implications for technology design. Although these studies are carried out largely in the context of giving material things (objects and money), there are other ways to "give," and we may find that there are benefits to the giving and receiving of praise, gratitude, and endorsement that goes on virtually. Many of us have experienced the warm fuzzies conferred by an opportunity to praise a friend on a social network or publically endorse a colleague for skills we admire.

Interestingly, these warm fuzzies also emerge when we virally share images and videos of people helping others at some cost to themselves (for example, the truck driver who rescued bear cubs from a trash can, the homeless man who found $40,000 and returned it, or any of the other spontaneous moments of altruism to which we deeply enjoy bearing witness). These acts are worth sharing because they *inspire* us, which leads us to another trigger for compassion.

Altruism and Inspiration—More Links to Wellbeing

If we attempt to explain altruism based purely on self-serving motivations, we find ourselves in a pickle when we come up against the fact that we are absolutely giddy at seeing other people do altruistic things. Jonathan Haidt (2005) uses the term *elevation* to describe the "warm, uplifting feeling that people experience when they see unexpected acts of human goodness, kindness, courage, or compassion. It makes a person want to help others and to become a better person himself or herself."

Research has also demonstrated the effects of elevation at work showing that self-sacrificing behavior among leaders can increase commitment and compassion among employees (Vianello, Galliani, & Haidt, 2010) and that elevation predicts volunteerism even three months after an inspiring experience (Cox, 2010). Inspiration carries with it a motivator to altruistic action that is separate from empathy. When we are inspired to act, we seem to be motivated by a renewed feeling of faith in human goodness and our ability to play a part.

Moreover, elevation seems to be a vector for the spread of altruism, which according to research can lead to a chain reaction within social networks (Fowler & Christakis, 2010). But it is the description of the heartfelt feelings evoked from one study participant that says it best: "I felt like jumping out of the car and hugging this guy. I felt like singing and running, or skipping and laughing. Just being active. I felt like saying nice things about people. Writing a beautiful poem or love song. Playing in the snow like a child. Telling everybody about his deed" (Haedt, 2005).

The links between compassion, altruism, and inspiration are also evident in the brain according to the work of Mary-Helen Immordino-Yang and her colleagues. Yang's neuroscientific research in this area (Immordino-Yang, McColl, Damasio, & Damasio, 2009) has shown that the neural circuits involved in interoceptive processing are activated in *both* admiration for virtue and compassion for another's pain. This work suggests that both states involve recognizing emotions and *reflecting on one's own behaviors* in response (Immordino-Yang, 2011). (See her sidebar in this chapter for further inspiration.)

If we know that bearing witness to altruism motivates kindness, compassion, and further altruism, why don't news programs spend more time on stories about compassionate people, altruistic action or progress as part of their lineup? How often are reports on current atrocities piled one against the other and perhaps capped off with a sport or beauty update? Many find the news overwhelming and avoid it, others watch without hope of being able to do anything in response, so think how much would change if news programs took the advice of empathy game-design researchers and ended reports with suggestions or depictions of related compassionate action? Recent initiatives such as the *Huffington Post*'s Good News section and the Good News Network have responded to the desire for positive news. However, these remain separate from the mainstream. We look forward to more balanced reporting in mainstream news that supports compassionate response and acknowledgement of progress instead of just bleeding leads.

You'll forgive us for dwelling so long on the research behind compassion, altruism, and its links to wellbeing, but without a sufficiently sophisticated grasp of these phenomena we would be little equipped to design genuinely helpful interventions to foster their development. Now, on to the interventions.

Interventions and Strategies for Cultivating Compassion and Altruism

Many short- and longer-term interventions for fostering compassion are based on meditation. Stefan Hoffmann, Paul Grossman, and Devon Hinton (2011) review intervention studies using compassion and loving-kindness meditation and conclude that they increase positive affect, lower negative affect, and can be effective in the management of many psychological problems, including depression, social anxiety, marital conflict, anger, and coping with the strains of long-term caregiving. Other studies have produced results showing that meditation-based compassion interventions increase not only compassion and prosocial behavior, but also overall psychological wellbeing.

What's interesting is the way in which compassion meditation does not dull one's emotional reaction to pain (to make it bearable) but instead results in an alternate response, specifically reducing amygdala activation associated with threat perception but increasing responsiveness to suffering (Desbordes et al., 2012; Lutz et al., 2008). This seems to support the idea that compassion is highly sensitive to suffering but responds with caring and approach rather than with distress or avoidance. In other words, in empathic distress we appraise another's suffering as threatening (to our sense of security perhaps), but when we feel compassion, we do not experience this threat response. Both of these studies also give evidence that the changes to neural activation that come as a result of compassion meditation can be enduring.

In compassion-focused therapy, Gilbert and colleagues have used multiple methods for eliciting and developing compassionate responses toward the self (see, e.g., Gilbert, 2010 and the edited volume, Gilbert, 2005). In order to foster feelings of warmth, they guide patients to bring to mind an image that, for them, represents their ideal of caring and compassion. In another intervention, patients write themselves compassionate letters. Psychologists also employ psychoeducation—that is, they explain to patients, using imagery, that just as images of food or sex are enough to elicit physiological responses of hunger or arousal, internal criticism is enough to induce the physiological stress and wear caused by real external criticism.

Research studies often use imagery, videos, and stories in a lab setting to induce states of compassion for experimental purposes, which suggests that these things might also be helpful in the context of technology environments, at least for priming or eliciting temporary compassionate response.

Technologies in the Development of Compassion

As mentioned previously, many of the approaches for fostering empathy discussed in the previous chapter are relevant to promoting compassion and altruism. After all, the purpose of games for change is not just to encourage people to empathize, but to inspire them to take action and affect change. Likewise, Jeremy Bailenson's work has given evidence for virtual reality's potential not only to increase empathy, but also to increase helping behavior and altruism (see his sidebar in this chapter).

In general, technology has been more interested in computer strengths than it has been in human ones. The ACM Digital Library lists 1,334 publications that contain the word *altruism*, and nearly all of them relate to computer altruism, which refers to collaborative behaviors in which a computer or software agent takes into account the interests of other computers. Excluding computer altruism leaves us with 198 studies on human cooperation, such as in business or for volunteer content generation—for example, wikis and open-source development. There are also some studies that apply the concept of altruism to new algorithms. However, very few involve the ways in which human altruism might be influenced by design.

Exceptions include a study (Davis, Farnham, & Jensen, 2002) that took what we would call a preventative design approach and explored the impact of different user-interface designs on how often players cooperated rather than "short-circuited" (or defected) a collaboration. Specifically, 25 pairs of participants were randomly assigned to using either an interface with text chat but no personal profile, a text chat but with a small personal profile, or a text-to-speech system with no personal profile. The results showed that the text-to-speech system (voice) reduced uncooperative behaviors, despite its being computer generated and gender neutral. The fact that this effect was stronger than having a profile page is quite significant and provides evidence that further work is needed to understand what drives prosocial and antisocial behavior in digital environments.

In a very different study, Yeoreum Lee and colleagues (2011) proposed a concept of "altruistic interaction" that requires participants to help and be helped. The participants designed and tested a fan that blows air on someone (produces an output) only when a person somewhere else blows into it (acts as input). In doing the blowing, participants felt they were helping another person, and the recipient was able to express gratitude. The design created a situation in which a user was able to help another, although issues of reciprocity and contingency blurred the line between altruism and a dependency that arguably compromised autonomy.

More familiar are examples within social media and social games. For example, in the farm game *Hay Day*, players can choose to help other players by tapping on fruit trees that have withered. Tapping revives a tree, allowing tree owners to grow fruit once again. It's entirely plausible that this small opportunity to help another, although only virtually, elicits some of the positive affect associated with giving.

Interestingly, however, the act eventually confers benefits on the performer in the form of thank-you certificates, a virtual currency that is *automatically* sent when the receiver accepts their revived tree. The result is a complicated mix of giving, receiving, and expectation of reward. When I (Dorian) played this game for the first time, I enjoyed the affirming glow of having been benevolent each time I revived someone's wilted tree. Once I got conditioned to the idea that I would receive payment, however, the glow diminished. I found that the opportunity for altruistic warmth had been replaced by the dopaminergic drive to accrue certificates, so familiar to gameplay.

The idea that anticipated payback might undermine the psychophysiological benefits of compassion and altruism sound suspiciously like the capacity for extrinsic rewards to undermine intrinsic motivation. Exploring a balance between opportunities to give *unconditionally* and opportunities for reciprocity and social exchange in games is a fascinating area for future research and experimental design.

Prosociality through Games

Although the relationship between violent games and behavior has received considerable research and media attention, an increasing number of developers and researchers are exploring the impact of games designed to support positive qualities and prosocial behaviors.

For example, in one study (Gentile et al., 2009) a team of researchers from Japan, Singapore, Malaysia, the Netherlands, and the United States reported the effects of prosocial games on the prosocial behaviors of people within three different age groups and across three different countries over extended periods of time. Critically, the study confirmed the hypothesis (controlling for sex, age, and time spent playing) that prosocial game exposure had a positive *causal* relationship to prosocial behaviors and traits. The study used careful experimental designs combining correlational and longitudinal analysis and converging results. The evidence showed that just as playing violent games produces hurtful behavior, playing prosocial games produces helpful behavior in both the short and longer term.

An interesting aspect of the study is how it demonstrates the way in which multiple instruments (i.e., questionnaires) can be used to measure aspects of prosocial behavior. The study combined the use of

- A prosocial orientation questionnaire that had participants rate statements such as "I would spend time and money to help those in need"
- The Children's Empathic Attitudes Questionnaire used to measure trait empathy through statements such as "When I see a student who is upset, it really bothers me"
- The Normative Beliefs about Aggression Scale (e.g., "In general, it is OK to hit other people"
- Stories with ambiguous provocative situations (e.g., someone scratches your car) to measure hostile attribution bias by asking participants to explain the situation ("he meant to scratch it")

Each of these instruments has been independently tested for reliability and validity to support their use in psychological research. More importantly for our purposes, designers can use them to measure the impact of their designs.

These studies used out-of-the-box commercial games. Given the evidence that prosocial (i.e., positive) games develop altruism and other prosocial behaviors (linked to wellbeing), we expect to see greater interest in these wellbeing-boosting games from parents, teachers, developers, and gamers.

Virtual Reality—Embodying Altruism and Helping Behavior

In a recent study published in *PLOS ONE*, a team at Stanford led by Jeremy Bailenson (Rosenberg, Baughman, & Bailenson, 2013) found that augmented virtual-reality games could lead to increases in altruism. Half of the 60 participants who completed the study were given the virtual power to fly like Superman (the "superhero" condition), while the other half could fly around the same space in a virtual helicopter. In the two-by-two design, participants in each of these groups were also allocated either to helping a sick child or touring a virtual city. At the end of the virtual-reality experience, participants were confronted by "someone in need of help" (an actor). The researchers measured the time to help and the amount of help provided by those in the different experimental conditions, and the results showed that those in the superhero/child-saving condition were significantly faster and helped more than those in the touring conditions. Six of the touring participants didn't help at all, whereas all of the former superheroes did. The researchers hypothesized that the embodied experience of helping facilitated the transfer of this behavior to the real world. As we have seen, prosocial games foster prosocial behaviors even with lower tech immersion, so the idea that embodiment might play a role in fostering compassion and altruism is unsurprising. These studies seem to suggest that giving people practice in helping (or perhaps the experience of being capable of helping—increasing their sense of coping ability) inspires prosociality even after the game is over.

Design Implications

According to evolutionary and emotion appraisal theories (Ortony, Clore, & Collins, 1988), we can conclude that the following conditions **increase the likelihood** of compassion arising:

1. **Relevance/similarity:** the target is perceived as belonging to the same family or group or is perceived as similar to oneself in another way (perhaps via a sense of shared humanity).
2. **Goal congruence/fairness:** the target could engage in future cooperation and is not to be blamed for his or her suffering.
3. **Empowerment/ability to cope:** the individual can cope with the cost of behaving in consequence to the compassion emotion (i.e., coping ability).

As a start, these factors (relevance, goal congruence, and coping ability) could be considered logical targets for systems seeking to develop compassion.

Digital systems that aim to connect people, to help them work together, or to allow them to help each other are in a unique position to leverage what we know about the factors that increase compassion and altruism. Take, for example, nonprofit organizations targeting poverty through microlending, such as Kiva and GoodReturn. They have two main user groups: low-income entrepreneurs who need funding for small ventures (e.g., a motorcycle to start a home-delivery service) and higher-income funders who wish to contribute to social good as microinvestors.

Over the past few years, I (Rafael) have been involved in a number of small projects on behalf of GoodReturn. Although we have yet to attempt to apply compassion research to these projects, we could, for example, attempt to match charitable projects to users based on compassion-appraisal factors. In the case of first-time donors, for example, a system that could match investors with entrepreneurs based on goal congruence (or even taking into account investor coping ability) may increase the effectiveness of these sorts of environments.

One interesting thing about microlending is that it includes evidence of responsibility or *deservedness*, which, as we have seen influences altruism positively. Those seeking funding are doing so in order to actively combat their own poverty, and they commit to repaying a loan. For situations involving charitable giving in contrast, appraisals of responsibility will be more prone to affect prosocial behavior, so designers may need to address perceptions regarding blame.

Design to Address Judgment and Blame

In light of the inhibiting effects that judgment has on compassion and altruism, designers seeking to foster social change will often need to address underlying perceptions having to do with issues of suffering. In some cases, this will be about correcting misconceptions (about, for example, the various roots of mental illness or poverty). In other circumstances, strategies for fostering empathy may be the first step, such as for conflict resolution (as with the *PeaceMaker* game).

By way of example, Belman and Flanagan (2010) give the hypothetical example of a game for eliciting empathy in relation to homelessness (in aid of soliciting aid). They point out that for players who attribute blame to those who are homeless, although engaging with the game may increase their experience of empathy, they are unlikely to give time or money to a shelter because their appraisal of the situation is incongruent with their goals. In other words, people's attitudes toward homelessness will affect whether the game leads to compassion and helping behavior or stops at empathy.

Finally, there is an open question as to how technologies can be used to foster the nonjudgmental aspect of compassion, based on a notion of common humanity.

Design for Inspiration

As we have seen, empathy and compassion aren't the only triggers for altruism. Based on the work by researchers such as Haidt and Immordino-Yang, allowing people to witness other people's compassion and perhaps supporting reflection on their own behavior could prove to be another effective strategy to cultivating compassion and altruism. Supporting the sharing of inspiring images, videos, and stories across social networks provides one simple example of inspiration sharing.

For other contexts, designers might use elevation as a way to support a sense of empowerment/coping ability as a way of transforming empathy into action. As mentioned previously, many charities have moved from depicting extreme suffering to depicting the fruits of helping (empowerment), suggesting that it is the latter that more consistently inspires giving. There are a number of exciting paths open to exploration in this area, and we expect to see a growing area of special interest at the intersection of inspiration, elevation, and technology over the next decade.

Detecting Compassion

Where technologies or games seek to foster empathy or compassion but do not have access to measuring consequential action, the question arises as to how technologists might automatically know if their efforts are at all helpful. As far as we are aware, compassion detection has yet to be attempted with affective computing (Calvo, D'Mello, Gratch, & Kappas, 2014), but research suggests there are physiological signals and facial expressions unique to compassion, so there is potential for noninvasive automatic techniques that would certainly prove helpful to work in this area.

Movement and Synchrony

According to fascinating research on dance, rituals, and movement, there is evidence that *synchronous movement* can enhance cooperative ability, compassion, and even altruism (Behrends, Müller, & Dziobek, 2012; Valdesolo & DeSteno, 2011). Although we have yet to come across videogames designed deliberately to leverage this intriguing phenomenon, the implications for physical systems such as the Wii or gesture-camera-based systems such as Kinect are self-evident. We eagerly await developments at this intersection of movement, compassion, and games. Until then, keep dancing.

In conclusion, in an article on the wellbeing benefits of compassion, Emma Seppala (2013), associate director of the Stanford Center for

Compassion and Altruism Research and Education, concludes that, "thanks to rigorous research on the benefits of compassion, we are moving toward a world in which the practice of compassion is understood to be as important for health as physical exercise and a healthful diet, empirically validated techniques for cultivating compassion are widely accessible, and the practice of compassion is taught and applied in schools, hospitals, prisons, the military, and beyond." As technologists, we can be part of making that world. Seeing as technology is now woven into the fabric of each of these areas, we'll need to be.

Expert Perspectives—Technology for Altruism and Inspiration

Virtual Altruism

Jeremy Bailenson, Stanford University

Imagine looking in the mirror and watching yourself transform to a different race or gender. At Stanford's Virtual Human Interaction Lab, my team and I have spent the past decade researching how virtual reality can create a feeling of empathy by allowing users to "walk a mile" in someone else's shoes and by using technology as a tool to reduce prejudice and increase altruism.

For example, Silicon Valley corporations are working with us to develop diversity-training software—instead of simply imagining a scenario related to racial discrimination or sexual harassment, learners can experience these

(continued)

> scenarios directly, receiving a visceral sense of the consequences of those behaviors. This technology can also be used to help people understand others in differing demographic populations. If in virtual reality you wear an avatar with a disability, our research has demonstrated that after the experience you are more willing to assist others with a disability, as compared to experimental participants in control conditions. Taking it up a notch, what if you could walk a mile in the shoes of someone who doesn't wear shoes—a chicken, an owl, or an octopus?! The lab's recent work has extended the notion of virtual empathy and altruism into the realm of environmental issues. When people are embodied in the avatar of a cow and then virtually prodded toward the slaughterhouse, will they continue to consume meat at every meal? If you're embodied as a fish in a coral reef and watch your home disintegrate due to pollution and ocean acidification, might this change your views toward product consumption that is harmful to the ocean? Virtual experiences feel real, and they allow for a user to experience the impossible.

(continued)

Developing Computer Interfaces That Inspire: Insights from Affective Neuroscience

Mary-Helen Immordino-Yang, Rossier School of Education, Brain and Creativity Institute, University of Southern California

Witnessing the exceptional achievements of others often leads to a subjective feeling of inspiration—a conscious desire to accomplish greater things oneself. Research suggests that inspiration varies depending on whether the action admired is remarkable for its moral implications (e.g., Martin Luther King Jr.'s activism) or for its skillfulness (e.g., Magic Johnson's basketball playing). According to our ongoing research at the University of Southern California, whereas reactions to skillfulness are relatively direct and cognitively concrete, reactions to virtuous acts often require more abstract cognitive processing, like the calling up of personal values and memories. Reactions to virtue also tend to be more generalized than reactions to skill. For instance, whereas one who admires Magic Johnson's playing may try harder during basketball practice, someone who admires Martin Luther King Jr. may apply her motivation more broadly—vowing to work harder toward accomplishing a goal relevant to her own personal beliefs and values, such as becoming the first in her family to graduate from college.

Our research suggests that morally and socially complex varieties of inspiration may involve distancing oneself from the current context in order to build connections to past experiences, personal values, and possible futures. This virtue-related inspiration might be particularly useful, if potentially more

(continued)

> difficult, to promote through positive computing. How might neurobiological insights help with this aim? In an ongoing series of studies, my team and I are exposing adolescents and young adults to emotion-provoking true stories, some of which are intended to be inspirational. Participants discuss the stories in a two-hour interview and then undergo fMRI scanning with psychophysiological recording as they review the stories and report their emotion to each in real time.
>
> These experiments reveal systematic connections between how people describe feelings of inspiration, their behaviors during these feelings, and neurobiological correlates. For example, just before participants report feeling inspired in the interview, they tend to avert their gaze away from distractions in the immediate environment, turning their eyes to a blank wall and incorporating long pauses into their speech. Increased gaze aversion correlates with increased reports of feeling inspired and with increased cognitive complexity and mentions of values. Eye-gaze aversion also correlates with individual differences in neural activity in a region known to be involved in controlling eye saccades and visual attention in monkeys as well as in personal memory processing in humans. These findings suggest that eye gaze may be a behavioral indicator of patterns important to inspiration.
>
> Another recent analysis suggests that both personal memory retrieval and visceral feeling mechanisms are active during reflective pauses, giving new insights into why reflective pauses may be important for emotional and social meaning making during learning. Although this research is in no way ready for direct introduction into computer software, it does provide a tantalizing suggestion that neuroscientific evidence could potentially be used to identify instances of effortful internal reflection during HCI, distinguishing reflection from daydreaming or loss of attention.
>
> It might eventually be possible for computer interfaces to use real-time readings of human eye-gaze patterns to adjust their responses in ways that would facilitate reflection on the digitally mediated experience. This innovation could potentially support learners and other users in building more complex conceptual understandings and in becoming more emotionally engaged with and possibly even inspired by what they are learning.

References

Aknin, L. B., Barrington-Leigh, C. P., Dunn, E. W., Helliwell, J. F., Burns, J., Biswas-Diener, R., ... Norton, M. I. (2013). Prosocial spending and well-being: Cross-cultural evidence for a psychological universal. *Journal of Personality and Social Psychology, 104*(4), 635–652. doi:10.1037/a0031578.

Aknin, L. B., Hamlin, J. K., & Dunn, E. W. (2012). Giving leads to happiness in young children. *PLOS ONE, 7*(6), e39211.

Batson, C. D. (1991). *The altruism question*. Hillsdale, NJ: Erlbaum.

Batson, C. D. (2002). Addressing the altruism question experimentally. In S. G. Post (Ed.), *Altruism and altruistic love: Science, philosophy, and religion in dialogue* (pp. 89–105). Oxford: Oxford University Press.

Behrends, A., Müller, S., & Dziobek, I. (2012). Moving in and out of synchrony: A concept for a new intervention fostering empathy through interactional movement and dance. *Arts in Psychotherapy, 39*(2), 107–116.

Belman, J., & Flanagan, M. (2010). Designing games to foster empathy. *Cognitive Technology, 14*(2), 5–15.

Brown, S. L., Nesse, R. M., Vinokur, A. D., & Smith, D. M. (2003). Providing social support may be more beneficial than receiving it: Results from a prospective study of mortality. *Psychological Science, 14*(4), 320–327.

Calvo, R. A., D'Mello, S. K., Gratch, J. & Kappas, A. (Eds.). (2014). *Handbook of affective computing*. New York: Oxford University Press.

Conrad, D., & Kellar-Guenther, Y. (2006). Compassion fatigue, burnout, and compassion satisfaction among Colorado child protection workers. *Child Abuse & Neglect, 30*(10), 1071–1080.

Cosley, B. J., McCoy, S. K., Saslow, L. R., & Epel, E. S. (2010). Is compassion for others stress buffering? Consequences of compassion and social support for physiological reactivity to stress. *Journal of Experimental Social Psychology, 46*(5), 816–823.

Cox, K. S. (2010). Elevation predicts domain-specific volunteerism 3 months later. *Journal of Positive Psychology, 5*(5), 333–341.

Davis, J. P., Farnham, S., & Jensen, C. (2002). Decreasing online "bad" behavior. In *CHI'02 extended abstracts on human factors in computing systems* (pp. 718–719). New York: ACM.

Desbordes, G., Negi, L. T., Pace, T. W. W., Wallace, B. A., Raison, C. L., & Schwartz, E. L. (2012). Effects of mindful-attention and compassion meditation training on amygdala response to emotional stimuli in an ordinary, non-meditative state. *Frontiers in Human Neuroscience, 6,* 292.

Dunn, E., & Norton, M. (2013). How to make giving feel good. Greater Good, the Science of a Meaningful Life—University of California, Berkeley, June. Retrieved from http://greatergood.berkeley.edu/article/item/how_to_make_giving_feel_good#.

Dunn, E. W., Aknin, L. B., & Norton, M. I. (2008). Spending money on others promotes happiness. *Science, 319*(5870), 1687–1688. Retrieved from http://www.ncbi.nlm.nih.gov/pubmed/18356530.

Elsass, P., Carlsson, J., & Husum, K. (2010). Spiritualitet som coping hos tibetanske torturoverlevere (Spirituality as coping in Tibetan torture survivors). *Ugeskrift for Laeger, 172*(2), 137–140.

Fehr, E., & Fischbacher, U. (2003). The nature of human altruism. *Nature, 425*(6960), 785–791.

Fowler, J. H., & Christakis, N. A. (2010). Cooperative behavior cascades in human social networks. *Proceedings of the National Academy of Sciences of the United States of America, 107*(12), 5334–5338.

Gentile, D. A., Anderson, C. A., Yukawa, S., Ihori, N., Saleem, M., Ming, L. K., ... L.K. Ming. (2009). The effects of prosocial video games on prosocial behaviors: International evidence from correlational, longitudinal, and experimental studies. *Personality and Social Psychology Bulletin, 35*(6), 752–763.

Gilbert, P. (2005). *Compassion: Conceptualisations, research, and use in psychotherapy.* London: Routledge.

Gilbert, P. (2010). An introduction to compassion focused therapy in cognitive behavior therapy. *International Journal of Cognitive Therapy, 3*(2), 97–112.

Gilbert, P. (2013). *Mindful compassion* (p. 384). London: Constable & Robinson.

Gilbert, P., & Procter, S. (2006). Compassionate mind training for people with high shame and self-criticism: Overview and pilot study of a group therapy approach. *Clinical Psychology & Psychotherapy, 379,* 353–379.

Goetz, J. L., Keltner, D., & Simon-Thomas, E. (2010). Compassion: An evolutionary analysis and empirical review. *Psychological Bulletin, 136*(3), 351–374.

Goleman, D. (2003). *Destructive emotions: How can we overcome them?* New York: Bantam Books.

Haedt, J. (2005). *Wired to be inspired.* Berkeley: Greater Good Science Center, University of California.

Hein, G., & Singer, T. (2008). I feel how you feel but not always: The empathic brain and its modulation. *Current Opinion in Neurobiology, 18*(2), 153–158.

His Holiness the 14th Dalai Lama of Tibet. (n.d.). Compassion and the individual. Retrieved from http://www.dalailama.com/messages/compassion, December 16, 2013.

Hofmann, S. G., Grossman, P., & Hinton, D. E. (2011). Loving-kindness and compassion meditation: Potential for psychological interventions. *Clinical Psychology Review*, *31*(7), 1126–1132.

Homans, G. C. (1958). Social behavior as exchange. *American Journal of Sociology*, *63*(6), 597–606. doi:10.1086/222355.

Immordino-Yang, M. H. (2011). Me, my "self," and you: Neuropsychological relations between social emotion, self-awareness, and morality. *Emotion Review*, *3*(3), 313–315.

Immordino-Yang, M. H., McColl, A., Damasio, H., & Damasio, A. (2009). Neural correlates of admiration and compassion. *Proceedings of the National Academy of Sciences of the United States of America*, *106*(19), 8021–8026.

Klimecki, O. M., Leiberg, S., Lamm, C., & Singer, T. (2013). Functional neural plasticity and associated changes in positive affect after compassion training. *Cerebral Cortex*, *23*(7), 1552–1561.

Klimecki, O. M., Leiberg, S., Ricard, M., & Singer, T. (2013). Differential pattern of functional brain plasticity after compassion and empathy training. *Social Cognitive and Affective Neuroscience*, nst060.

Kornfield, J. (2009). *The wise heart: A guide to the universal teachings of Buddhist psychology* (p. 448). New York: Bantam.

Kornfield, J. (2011). *Heart of forgiveness*. Berkeley: Greater Good Science Center, University of California.

Lee, Y., Lim Y., & Suk, H. (2011). Altruistic interaction design: A new interaction design approach for making people care more about others. In *DPPI '11 proceedings of the 2011 Conference on Designing Pleasurable Products and Interfaces* (pp. 59–62). New York: ACM.

Lutz, A., Brefczynski-Lewis, J., Johnstone, T., & Davidson, R. J. (2008). Regulation of the neural circuitry of emotion by compassion meditation: Effects of meditative expertise. *PLOS ONE*, *3*(3), e1897.

Neff, K. D. (2011). Self-compassion, self-esteem, and well-being. *Social and Personality Psychology Compass*, *5*(1), 1–12.

Neff, K. D., Kirkpatrick, K. L., & Rude, S. S. (2007). Self-compassion and adaptive psychological functioning. *Journal of Research in Personality*, *41*(1), 139–154.

Okun, M. A., Yeung, E. W., & Brown, S. (2013). Volunteering by older adults and risk of mortality: A meta-analysis. *Psychology and Aging*, *28*(2), 564–577.

Ortony, A., Clore, G. L., & Collins, A. (1988). *The cognitive structure of emotions.* Cambridge, UK: Cambridge University Press.

Reyes, D. M. (2011). Self-compassion: A concept analysis. *Journal of Holistic Nursing, 30*(October), 81–89.

Ricard, M. (2007). *Happiness: A guide to developing life's most important skill* (p. 304). New York: Little, Brown.

Rosenberg, R. S., Baughman, S. L., & Bailenson, J. N. (2013). Virtual superheroes: Using superpowers in virtual reality to encourage prosocial behavior. *PLOS ONE, 8*(1), e55003.

Seppala, E. (2013). The compassionate mind. *The Observer, Association for Psychological Science, 26*(5), 20–29.

Thomas, J. (2013). Association of personal distress with burnout, compassion fatigue, and compassion satisfaction among clinical social workers. *Journal of Social Service Research, 39*(3), 365–379.

Valdesolo, P., & DeSteno, D. (2011). Synchrony and the social tuning of compassion. *Emotion, 11*(2), 262–266.

Vianello, M., Galliani, E. M., & Haidt, J. (2010). Elevation at work: The effects of leaders' moral excellence. *Journal of Positive Psychology, 5*(5), 390–411.

Wispé, L. (1986). The distinction between sympathy and empathy: To call forth a concept, a word is needed. *Journal of Personality and Social Psychology, 50*(2), 314–321.

12 Caveats, Considerations, and the Way Ahead

Ten years ago, if you'd told me (Dorian) that technology would get involved in things as personal as mindfulness, happiness, and wellbeing, I would have covered my ears and said, "Make it go away." In fact, that's kind of what I did when Rafael first proposed the antecedents of this book. Wasn't technology already part of the problem? Didn't it already get in the way of happiness and intrude far too much into our humanity? People are losing touch with nature, with the tactile, the alive, the private, the unmediated, and they're forgetting how to focus, introspect, or engage in reflection that isn't subsequently performed as a status update. We seem to be increasingly disconnected from the present moment, from the raw reality of the physical world and from our friends and families *because* of technology. And now we want it to invade the few spaces we have left to keep us sane? Despite the fact that I'm far from a Luddite and have worked as a digital designer since the 1990s, I have to admit that these thoughts have often run through my head.

However, as time passed, I began to consider a few things. First, technology already was *intruding* into these areas, and if that intrusion was in fact detrimental, there should be a way of proving it and changing it. Second, perhaps it wasn't technology per se, but the way we'd been designing and using it that was the problem, and perhaps we could do better.

In the past few years, we've had the privilege of coming across researchers and practitioners doing absolutely inspiring work. Some of them have had the incredible generosity to share their ideas with us for this book. Examples of technology that saves lives through education, cultivates compassion through role play, enriches lives by offering new forms of creation and meaning making, connects grandparents with grandchildren, allows those with disabilities to be empowered as never before, curbs anxiety, reduces depression, and opens the door to positive change—examples of deep connection, engagement, and meaning that were never possible before—have made me see that technology is not the problem any more than it is the solution. It can support wonderfully life-affirming things or perpetuate our own destructive habits, and which direction it goes will always be up to us—the people who design and consume it.

Nevertheless, we will always maintain that not everything in our lives should be mediated by technology. There will be times when the best thing for someone's wellbeing will be to shut down her computer and go on a retreat, tend to some vegetables in the garden, play a game of real soccer on real grass, or have a get-together with friends without the presence of mobile phones. There will be times at work when, in order to focus on your writing, stimulate creative thinking, or connect meaningfully with the people around you, you will be better served by the absence of devices, by a pencil and paper, by hands and faces. We tend to get caught up in the idea that ubiquity and pervasiveness are inevitable and desirable, without stopping to question those assumptions. Maybe a vast increase in embedded technologies will make everyone happier, but can we assume that it will? A march toward ubiquity could just as believably infringe not only on our privacy, but also on our autonomy and wellbeing. In other words, is there a point at which it no longer makes sense to continue to add digital technology to everything? Do the benefits outweigh the costs, and are we appropriately measuring the costs?

As technologists, we have to learn how to stand by the radical notion that adding technology won't always make experience better and that the answer to making something better won't always be technology. Once we can do this, we can focus our efforts and resources on those places where it really *can* make a significant difference.

In this final chapter, we take a moment to honor the complexity of human experience and propose a humble, balanced approach to technology development. We attempt to "out" or shine an early spotlight on some of the various pitfalls and misuses that may quietly emerge as we seek to design for wellbeing. We have elected to bring attention to these issues not because we have clever solutions to them, but because they are open questions that we believe should not be neglected in our consideration, research, and ongoing debate moving forward. We close with a decidedly pragmatic look at a way forward for positive computing, both as a field of research and as an area of development that will help us to shape the future of our world.

Humans as Complex and Inconsistent Creatures

As technologists, we are keenly aware of how enthusiastic our community can be about technology intervention. When a new algorithm, technique, or design innovation is released, we have a tendency to want to apply it to anything and everything. We all have found ourselves asking how we might apply a newfound skill set, an approach we're familiar with, or our current abilities to something new. With the best of intentions, we carry around a solution in search for a need. This will always be part of the way that technology and research move forward. But we are also keenly aware of the downsides to this approach. When technologies are imposed on users, when inappropriate solutions are imposed on problems, new problems arise.

As we have alluded to previously, there is a tension between (*a*) the reality of humans, their minds, and their wellbeing as things that are incredibly complex, variable, and individual, and (*b*) the need for designers and developers to seek out generalizable models and manageable goals and to operationalize abstract notions into concrete solutions. Some level of reductionism will always apply. The danger is obvious: we can oversimplify and end up doing harm rather than good.

Take, for example, the quantifying of self. Whereas person A is inspired by the progress she sees in her tracked weight loss and exercise regimen, person B is angry that her scale is telling her she's not good enough and eats more sweets to compensate, and person C uses the tracking to reinforce already unhealthy attitudes and fuel excessive weight loss, self-criticism, and pathology. Humans are complicated. Many admonish friends or family for overindulging in cigarettes, calories, or beer, thinking that reproach is a sensible thing to do, when really it's fueling rebellious behavior, shame, or depression, which leads to more of the problem. Humans are complicated.

We have already alluded to what psychologists call "ironic processes" and the mind's ability to go against its own intentions. If I tell you that for the next 30 seconds, you should *not* think about food, you'll think about it more.

We don't believe that the idea that some people will always use technology in unintended ways is a reason not to develop it. However, as is true in other design contexts, being aware of edge cases and extreme users should help inform the way we design, distribute, and contextualize our innovations and prevent us from falling into the trap of oversimplifying our models of human thinking. Moreover, our measures for how well a technology is working and for whom it is most usefully targeted need to include evaluation over the longer term and employ multiple measures. Because people are complicated.

Privacy and Security

At the time we were writing this book, secret documents from the National Security Agency were leaked that revealed how personal communications by individuals around the world are recorded and data mined by security agencies and by the consulting companies that work for them, who allege they are doing so for security reasons. With no transparency, it is impossible to say what is legitimate. Whereas police officers need a warrant to search your house, those with access to our private conversations and activity traces thus far require no warrant to search them. Even if this process were transparent and justified, that the data exist and are entrusted to private companies means they're subject to commercial misuse and corruption internally as well as to data theft and terrorism externally.

Closer to our everyday lives is the way these data are already legally mined for commercial purposes. Most of us realize that private companies such as Facebook, Google, and Microsoft mine the data we provide (via web searches, social media interaction, etc.), so they can use what they learn about us to sell us products more effectively, which is why they can provide services for free.

As free-software evangelist Richard Stallman would say, these services are free as in beer, not free as in freedom. Product marketing involves attempts at behavior change, which is why marketers are among those who employ behavior design and persuasive strategies. Advertisements, by nature, exploit our ways of thinking and feeling in order to convince us to act or buy. In exactly the same way, these mined data could be used to inform propaganda to incite subtle changes that may even be in line with what is understood to improve our health, our wealth, or even our psychological wellbeing. It is our responsibility to remain critical of any behavior change incited subconsciously (untransparently) and with the use of personal data. We should always question the motivations and ethical drivers behind the organizations and governments with whom we trust knowledge about our lives.

Our position is that positive-computing shouldn't be used as an excuse to further encroach on our privacy—in other words, much can be done with the data already collected transparently by software systems—and that we need to be wary of rigid prescriptions for change. We should support transparency, autonomy, and conscious engagement in any effort for positive change. We hope some of the examples given in this book have reflected this notion, and we hope that designers and researchers will remain sensitive to these issues as we work to appropriately manage and keep control of our privacy, security, and autonomy as data spread out and into proprietary clouds.

Replacing Humans—Livelihood and Wellbeing

At a recent research summit, one of the talks included a graph evaluating the accuracy of computer vision techniques: humans, 98 percent; algorithm, 93.5 percent. "We are only 4.5 percent behind!" the speaker exclaimed enthusiastically, ironically positioning us all in the algorithm column rather than in the human column. Her talk is just a reflection of the larger view that those of us in computer science and engineering often take, competing with our own species as a way of setting targets for progress.

Most of us harbor hopes that technology will continue to surpass our skills, thereby serving and augmenting our capabilities as it has so many times before. After all, it is by applying technology that we are now able to travel at unnatural speeds, send messages over thousands of miles in nanoseconds, and fly over oceans. It is thanks to engineering efforts that a 100-ton truck can be driven into an Australian mine—where most humans dare not go owing to the blistering heat, collapsing rock, deadly snakes, and hairy spiders—by someone sitting in the comfort of an air-conditioned office in Sydney. Those of us who spent childhood hours watching *The Jetsons* dreamed of having a Rosie to sweep and cook and one of those conveyor belts that gets you showered and dressed in the morning. Technologies have long been designed to do those things we have done ineffectively, too expensively, or unenthusiastically.

But the trade-offs are not always straightforward. Almost 20 years ago, I (Rafael) worked as a consultant for a government project training young and unemployed Argentineans in a trade. I'll never forget the glowing look of pride on the faces of supermarket checkout trainees when they were praised for their newly acquired customer-service or procedural skills. They were gaining autonomy and competence. It is opportunities like these, the ones available to a wide group of people, that are increasingly replaced by machines.

Likewise, the driverless car, which we mentioned previously, no doubt will open new worlds of opportunity for some, but the millions more whose working lives depend entirely on transporting people and freight in taxis, buses, and trucks across countries and continents are facing a devastating threat to their livelihoods.

Of course, technology has always led to shifts in livelihoods, and some would argue these new technologies will decrease certain types of work but increase knowledge jobs. However, it's hard to imagine that trade-off will be one to one. If the point of new technology is often to reduce costs (not simply move those costs around), can we really hope to replace one million trade jobs with one million "knowledge" positions? Perhaps, but the argument also makes a value assumption that an office-based knowledge job is superior in everyone's view to other kinds of work. Driving a truck is not like working in the life-threatening depths of a mine. Many truckers value the autonomy and tranquility of their job on the road. Assuming that everyone has the secret wish (and appropriate capacity) to "upgrade" their profession feels like the insular view of a tech industry little connected to the people affected by its progress. Although it's true that over history we have been able to adapt to these changes and have generally ended up better off, there is no reason to assume that this proposition has no limits whatsoever to its advantage.

In fact, Erik Brynjolfsson at MIT's Sloan School of Management has identified an emerging paradox: "Productivity is at record levels, innovation has never been faster, and yet at the same time, we have a falling median income and we have fewer jobs" (Rotman, 2013). In their book *Race against the Machine*, Brynjolfsson and coauthor Andrew McAfee (2011) argue that this technological revolution, driven by advances in artificial intelligence and increased computing power, rather than eliminating blue-collar jobs, as has happened in the past, is now eliminating white-collar jobs, such as clerical work, accounting, and those that generally support the middle class. This has shrunk the middle class and increased economic inequality, something shown to decrease psychological wellbeing.

There are other reasons why replacing humans would be foolhardy, and we have mentioned some of them with regard to technologies for mental health. There are things people are better at, in both obvious and quantifiable ways, but also in less obvious qualitative ways, because human presence and connection are important too. Sure, we might be able to get on without a hello from the banker, local advice from a taxi driver, or a chat with the grocer, but at what point will this loss of human contact begin to disconnect and isolate us? Connection to others, to a community, and to a sense of common humanity has been shown to be critical to happiness, compassion, self-compassion, and many of the other aspects of wellbeing we have discussed in this book. To what extent do we prioritize efficiency, convenience, and profit for a few at the expense of opportunity, autonomy, and wellbeing for many?

A full discussion of this issue is, of course, far beyond the scope of this volume, but we bring it up because it's relevant to wellbeing. After all, building a technology that causes some measurable increase of wellbeing to users but contributes to the loss of livelihood for many more farther down the chain just might negate its positive effect. So social and economic impact is something researchers in positive computing should take into account. The issue is not so much about saying that no one should build machines that replace humans, but about questioning our assumptions about benefit and value. It is about using these considerations to help guide our focus and our resources toward work that can have benefit to wellbeing from a holistic perspective.

Who's in Control? Autonomy, Competence, and Empowerment

In a stirring scene from the biopic *Ray* (Taylor Hackford, 2004), musician Ray Charles as a very little boy falls to the ground, apparently helpless. Already completely blind, he instinctively cries out to his mother to come and help him. In this poignant moment, his mother, who stands only a few feet away, makes a powerful and courageous decision not to help him. Forced to help himself, he soon finds his way out of his predicament, and this moment of transition symbolizes the beginning of an independence—stubborn in the face of both physical disability and racism—that would be critical to his success in life.

All parents must navigate a border between providing support and encouraging the development of independence in children, and the issue is all the more complex at those various times in our lives, whether young or old, injured or ill, when we require assistance or adaptation in the face of change. Technology has the ability to assist in ways that either support or undermine our autonomy.

A current project we are involved in provides a uniquely illustrative example: How might technology be used to support young people with chronic illness in the transition from pediatric to regular care? For children growing up with chronic illnesses such as diabetes or cystic fibrosis, parents will be responsible for taking them to appointments, tracking medication compliance, and otherwise managing the illness as caretakers. However, there comes a time when they transition from being children to being adults and must take full responsibility and have full control over their own care if they are able. Although it would be fairly straightforward to develop a medication reminder service or automated doctor appointment app for the mobile phone, and we could have added contingent rewards such as badges or virtual currencies to motivate compliance, these things may be convenient in the short term, but they do not at all address the core issue, which is developing and supporting autonomy.

In contrast, by looking at ways in which we can support these young adults in integrating self-care actions with their goals and values, visualizing the impact of their decisions, encouraging a growing sense of competence and autonomy, perhaps in concert with a fading scaffold of practical support, we can begin to think more deeply not just about assisting, compensating, or training in the limited carrot-and-stick sense, but about respecting, empowering, and building opportunities for self-determination.

Owing to the fact that autonomy is a key component to wellbeing, the core of SDT, and validated cross-culturally, making a concerted effort to respect user autonomy and design for it is all the more critical to positive computing. Unlike other factors such as self-awareness and compassion, design decisions about autonomy are automatically embedded into all the technology we create.

Batya Freidman (1996) brought attention to user autonomy in the context of her seminal paper on VSD: "Autonomy is protected when users are given control over the right things at the right time. Of course, the hard work of design is to decide these what's and when's." She goes on to identify aspects of systems that can promote or undermine user autonomy, including system capability, system complexity, misrepresentation of the system, and system fluidity. Though this article was written almost 20 years ago now, we seem to persist in undervaluing the importance of autonomy, both at the software task level and at higher levels of human experience.

More recently, in *The Design of Future Things* Don Norman (2007) discusses at length the problems associated with "inappropriate automation" and points to the multifaceted nature of autonomy and automation issues: "Automation is a system issue, changing the way work is done, restructuring jobs, shifting the required tasks from one portion of the population to another and, in many cases, eliminating the need for some functions and adding others." Assisting, helping, compensating, removing tasks, meeting needs—these are generally thought to be good things, but a closer look reveals that the design decisions we make on behalf of meeting these goals resonate in unexpected ways. In attempting to solve a problem technically, simplify a task, or compensate for one isolated need, we can lose track of the larger ecosystem at the expense of human agency and wellbeing.

By automating everything, users have no capacity to adapt, customize, do parts of tasks themselves, their way—they can be left feeling helpless, especially when the technology breaks down. In this view, users must adapt to the technology rather than be able to adapt the technology to themselves.

Most recently, Yvonne Rogers and Gary Marsden (2013) shed critical light on the history of HCI approaches to design for helping "those in need." They contest that in HCI we are frequently guilty of taking a third-person view that reflects an unequal partnership. In designing to compensate or "augment frailty," well-meaning designers who identify needs from the perspective of their own contexts perpetuate an "asymmetrical relationship between those who have and those who have not, underlying an uneasy dependency between those who need and those who can help."

Speaking from extensive experience designing assisted-living technology, and based on work for the developing world, Rogers and Marsden propose a new approach and a new rhetoric of *empowerment* that seeks to engage users in creating technology for themselves on their own terms. They advise that we invest more effort in design tool kits and design education to empower these users to be the creators of their own solutions. What Rogers and Marsden suggest is really the ultimate in autonomy. As researchers and developers in positive computing, we should always be questioning our approaches, lest we attempt to cultivate one aspect of wellbeing at the expense of another.

Who's in Control? Motives, Power, and Paternalism

There are at least two loci of control in the making of any technology: the user and the designer. Behind the designer, or design team, is an organization, a project, or a group of stakeholders with their own motives, values, and goals. Judgments made by designers and organizations about what should go into a technology and how it should be designed will always be shaped by how they expect to profit and by their personal and cultural perspectives. Given the impact that technologies can have on people's behaviors, on their ways of thinking and feeling, it is important to make these motives and values as explicit as possible. Designers should be candid about their motivations; they won't always be fully aware of them, so they should engage in practices that help them to become conscious of these motives.

This is true even when developers are government or nonprofit organizations. Even when intentions are utterly blameless and totally laudable, the values and expectations underlying the team have an influence. What will funding agencies expect to see in terms of outcomes? What do designers know about the user groups? As is evident in reactions to persuasive computing and nudging, many have valid fears about the makers of technology using it to manipulate people. Whether it's persuading for marketing or paternalistic care in the case of government-led wellbeing campaigns, there will be an ongoing tension between support and control. Sensitive as we are in the United States to the values and influence of government, only a small number of us question the values and influence of technology makers, simply because these makers largely share the same culture as we do. But many more outside of the United States are keenly aware of the fact that a relatively small number of wealthy Americans in Silicon Valley determine the design, distribution, and sale of digital technology based on their own cultural and socioeconomic context and set of values. It's important to acknowledge this concern about a cultural imperialism.

We must seek to be aware of the assumptions we make and the biases we carry and the fact that we base so many decisions on personal experience or limited testing. We also need to humbly acknowledge that US culture currently has disproportionate influence globally and that it is therefore everyone's responsibility in every country in our increasingly globalized society to not be insular, to acknowledge our biases, and to seek to think and act more broadly and empathically. How exactly? Fortunately, over the past 10 years, researchers in VSD have been developing approaches for making designer and user values explicit. For all these reasons and more, we strongly support the application of VSD to positive-computing work.

Well-Washing—Research versus the Bandwagon

Transparency with regard to motives and values will also help us to discriminate genuine cross-disciplinary research-based work from projects with weaker foundations and those largely hijacking a notion of wellbeing for profit or bandwagon effects. This bandwagon phenomenon is nothing new. We have watched genuine environmental concern and innovative green design precede a rush of "greenwashing"—a marketing strategy that combines unfounded environmental claims and deceptive design to exploit the public interest in living sustainably. Overuse of the words *natural* and

Caveats, Considerations, and the Way Ahead 267

organic without certification—even the superfluous addition of safer ingredients (such as baking soda to corrosive cleaners) is used to falsely convey a sense to customers that this product is somehow safer, healthier, or more sustainable. It's easy to see how wholesome concerns can be exploited for profit, and wellbeing is not immune to such exploitation.

The public's embrace of the new science of wellbeing has led to an unsurprising explosion of products and promises you might call "well-washing." Dorian's father, who worked in advertising for many years, often said, "If you want to sell me something, it has to make me money, get me fed, or get me laid" (with corresponding unsavory hand gestures that we fortunately can't replicate in a book). Interestingly, nowadays we can add "make me happy" to the list. It sounds wholesome enough, but the trouble is that companies that sell everything from shampoo to Coca-Cola are hijacking notions of wellbeing to sell products. The irony in the case of Coca-Cola is particularly thick—a company associated with both human rights abuses against workers and a role in the obesity epidemic has, at the time of writing, a section on its website that quotes Aristotle and Ghandi on "happiness." It's interesting that it cites wellbeing experts who can't object. The section carries on by "blinding with science" and providing a list of tips to be happy, a few of which are well founded, and one of which, of course, is to drink the company's product.

How then do we "keep it real"? Obviously there is no easy answer, but acknowledging motives will be critical. Just because we are privileging an apparently altruistic goal in working in the area of positive computing, that doesn't mean there won't always be other pressures, values, and ideas about how to pursue that goal.

As always, science—in combination with a healthy skepticism—is our greatest ally in the battle against snake-oil salesmen. We can't ensure the scientific method is upheld at the grocery store or in web marketing, but when it comes to our own work, we can require evidence-based approaches and protocols. In addition to evidence-based practice, perhaps our greatest prophylactics are multidisciplinary collaboration, multidimensional evaluation, and, of course, that design stalwart, iteration. In summary, we must:

- Be honest and explicit about our motives and values.
- Demand research integrity and scientific method.
- Ensure multidisciplinary collaboration.
- Employ multidimensional evaluations.
- Take an iterative approach that allows adaptation based on evaluation.

Wellbeing and Culture

Evidence thus far shows that the factors of wellbeing we have discussed in this book are beneficial to all humans across cultures. For example, there is no evidence that cultivating compassion or practicing mindfulness will help people on one continent but not on another. In addition, there is evidence for the cross-cultural validity of factors such as autonomy to human psychological wellbeing. *However*, the extent to which cultivating each is beneficial and, even more so, the extent to which various *strategies* for wellbeing will be most effective *are* influenced by culture.

For example, Nancy Sin and Sonja Lyubomirsky (2009) found that the most commonly used positive-psychology interventions (PPIs), which were developed by psychologists in Western countries, were most effective for people in those countries. "Members of individualist cultures, whose values and cultural perspectives are highly supportive of the pursuit of individual happiness, have been found to benefit more from PPIs than members of collectivist cultures. As a result, clinicians are advised to consider a client's cultural background, as well as his or her unique inclinations, when implementing PPIs. For instance, a client from a collectivist culture may experience greater boosts in wellbeing when practicing prosocial and other-focused activities (e.g., performing acts of kindness, writing a letter of gratitude), compared with individual-focused activities (e.g., reflecting on personal strengths)."

We can't also help but wonder if those in individualist cultures would benefit more from other-focused practices in the longer term and vice versa for collectivist cultures benefiting from individualist practices. The point is that culture does have some effect on how wellbeing factors are communicated, combined, and manifest.

Aside from strategy development, culture may influence how scales of measurement are designed and communicated. Researchers and psychologists working to evaluate and improve wellbeing among Tibetan torture survivors (Elsass, Carlsson, & Husum, 2010) found that although their Western methods were helpful, the Tibetans found Western conceptualizations of emotions and wellbeing to be unsophisticated. Specifically, the authors report that Tibetan leaders in interview "questioned the validity of our western rating scales and explained that our results might be influenced by the Tibetan culture, which among other things can be characterized as having a view and articulation of suffering much more complex than the units of our study's rating scales."

Caveats, Considerations, and the Way Ahead

Culture as a mediator of wellbeing and cross-cultural validation of instruments constitute ongoing work in social psychology and other fields. Although there is plenty of evidence for a degree of universality when it comes to what contributes to wellbeing, evidence also tells us that there is great value in seeking out cross-cultural collaborations in our work. We still have much to learn from each other. (See the work of Ed Diener, Shigehiro Oishi, and their colleagues for more insight into how culture and wellbeing interact [e.g., Diener & Diener, 2009; Schimmack, Radhakrishnan, Oishi, Dzokoto, & Ahadi, 2002].)

Balance and the Mean—When More Isn't Better

Most factors of wellbeing, wholesome though they may be, like oxygen and water can still be overconsumed. When it comes to positive emotions, reflection, engagement, and so on, there are situations in which *more* ceases to improve wellbeing. It's important we take this into account, lest we carry on down a naive if convenient path that assumes anything we do to increase any wellbeing factor can only be good. This middle path is evident in Richard Davidson's (2012) framework of emotional style, which includes six neurophysiologically distinguished bipolar scales pertaining to traits such as resilience, self-awareness, and attention. Each of these traits in excess can have negative consequences.

Of course, since we're talking about *positive* states, traits, and practices, it seems reasonable to assume it's much *harder* to overdo it with them (as with drinking water) than it is to overdo it with neutral behaviors (eating food) or negative behaviors (drinking alcohol). Therefore, the overriding message for technologists seems to be one of generous moderation or even cautious indulgence. It would probably be difficult to increase compassion to any detrimental point, but certainly in fostering empathy we might inadvertently foster distress. It's easy to imagine how practices for self-awareness could lead to self-absorption and self-centeredness, and it has been studied how engagement and flow can spill over into addiction.

Defining the tipping points (and methods for detecting these points) will be an ongoing area for investigation. In the meantime, we can tread more safely in the following ways:

- Monitor the longer-term effects of our interventions.
- Embrace an iterative process that evaluates and adapts to feedback and changes in user behavior.

- Consider ways in which support for increasing one factor can be kept in balance by support for another. For example, extrapolating from research findings in psychology, we may find that efforts to foster other-focused factors such as compassion balance out self-focused factors such as self-awareness. Mindfulness may balance out reflection and goal setting, and mindfulness of negative emotion may balance out support for positive emotion.

The Ecology of Wellbeing—Taking a Holistic View

Although we can only ever be dealing with a thin sliver of those things that influence the wellbeing of people through technology, if our overarching goal is to work on behalf of a social good, we must also consider social responsibility, or the ecology of wellbeing that exists around our development work. For example, if we are promoting hardware and devices, then we should consider a device's full life cycle and periphery of effects to ensure it's in congruence with wellbeing goals. After all, what good is a wellness device made in a sweatshop? To this end, we envision multidisciplinary projects that connect different levels of wellbeing-related efforts. For example, a mobile-device project might combine positive-computing design methods and VSD with sustainable industrial design and approaches to HCI for development. Sure, this is ambitious considering we have yet to get slavery out of the computer supply chain,[1] but then we did get to the moon, and when it comes to the wellbeing of our species and planet, we think it's safe to say failure is not an option. A movement toward holistic multidisciplinary approaches has the greatest promise in promoting flourishing from all its angles.

We've presented just a few of the ethical dilemmas and issues requiring care associated with doing research on technology for wellbeing, but, naturally, others will have occurred to you as you read through these pages. Each of us, from the unique context of our own professional backgrounds and personal experiences, will be positioned to bring new issues to light. We urge you to do so. Enthusiasm is essential to the energy that drives discovery, but when unbridled it can threaten the credibility of its cause. Happily, the many examples of inspiring work to which we've had the privilege of referring in this book are just the tip of an iceberg—we look forward to the growth of this enthusiastic and broadminded research community driven by a desire to make a better future.

Caveats, Considerations, and the Way Ahead

The Way Ahead

People talk a great deal about disruption. In our profession, disruption is proof you've had an impact—that your work mattered and that you've wielded power over the masses. Disruption also tends to be a bit of a cash cow, which adds to its appeal. However, we suspect that the greatest successes in positive computing probably won't be ushered into the world screaming "Game change!" from every social media corner. They won't necessarily involve a tsunami of sudden consumption. They might, however, arise gradually and over time as a result of the persistent and passionate work of many, quietly making reparations in the background, slowly changing the shape of experience, softening corners, removing problems until users forget they ever existed, and adding opportunities for kindness, for attention, and for connection until users forget they weren't there before.

But, of course, even quiet revolutions need funding. The research and academic work we have focused on over the chapters in this book are only one slice of a big pie. For a field of positive computing to flourish and be relevant, we need a broad ecology of stakeholders that includes not only designers, developers, and scientists, but also the policymakers, investors, managers, and entrepreneurs who will bring positive-computing visions to life for the public.

So in place of a long-winded, if heartfelt, rainbow pep talk of a conclusion, we close up with a no-nonsense proposition: How do we fund our work in positive computing and, critically, do so without compromising on human-centeredness?

Informed by a range of both commercial and noncommercial projects that already exist, we introduce here a number of funding models and pathways to economically sustainable positive-computing development that we have witnessed in action today. We discuss them briefly, along with some of the advantages and challenges associated with each in hopes that an overview will help lightly grease the pathway to action for those who want to get this party started (or see it grow).

Funding Positive Computing

Positive Computing as a Science
Until recently the most significant group of "stakeholders" for projects relating to positive computing has been researchers at universities or at

publicly funded research organizations that study overlaps between technology, psychology, and wellness. Ongoing work has occurred in psychology as Internet-based mental health promotion, as positive technologies in cyberpsychology, and as a wealth of work within HCI under various umbrellas (see, for example, the "mind and spirit" sessions at the Computer–Human Interaction Conference, the workshops on interaction design for emotional wellbeing, and special journal issues on related themes). Each of these groups is influenced by its unique research interests and culture.

Work in research is by far the best suited to form a solid foundation for wellbeing technology because research work must be validated via established measures, academic peer review, and replication. This process guarantees that, over time, we can accumulate knowledge about design factors that positively influence wellbeing for different groups of people. Academic and scientific research, relatively free from commercial interests, has the advantage of reliability and is well suited to higher-level questions and basic research.

However, the process is a slow and expensive one. Researchers at universities and research organizations are generally funded by national agencies for a specific amount of time (usually one to four years) to run a specific study. More than a year can pass between the point at which an idea is fleshed out and a project launches. Funding-agency policy often requires adherence to the original plan, yet implementation plans entailing technology can easily become outmoded in six months' time. As such, these kinds of projects are best suited to exploring aspects more persistent or universal than technological implementation details usually are.

Positive Computing as a Commercial Industry
When it comes to innovating at a faster pace and with finer granularity and in ways that are focused on immediate application, industry has the advantage. Commercial products for personal development have been popular in the West since at least the 1970s, particularly in the form of self-help books. The validation of such behavioral bibliotherapies was patchy back in the 1970s (Glasgow & Rosen, 1978), and it would seem that not much evaluation has occurred since then. Mobile technology has made the concept digital, and many of the apps we have already mentioned from large companies and from a significant number of start-ups are commercializing products aimed at helping people track and improve their physical and sometimes emotional wellness. According to a report by the Pew Research Center (Fox & Duggan, 2013), 69 percent of US adults track their

(or a loved one's) health, 21 percent of them using some form of technology (e.g., smartphones and sensors). There seems sufficient evidence to assume there is a profitable interest in these types of technologies among consumers and the companies that are investing in them.

Of course, the allure of wellbeing technology as a potential goldmine is not without its dangers. Among today's most familiar models of technology product is the "free" service. In this scenario users are not so much the customers as the "products." Advertisers and businesses able to profit from users' profile information pay for the service. At the end of the day, a company needs to generate revenues and so "monetize" their users.

The monetization of users is rather different from the relationship between, say, a therapist and his clients or a doctor and her patients. In contrast, both these types of relationships and the motives that drive them are better understood by those involved and are even regulated by legislation in order to protect patients from malpractice. If technology companies get involved in our wellbeing, who will protect us from profit-based misconduct? Are there viable models for a positive-computing company, or will positive computing in industry simply fade into more forms of marketing?

Positive Computing as a Public Good

Government organizations as well as nonprofit organizations funded by private or public means have a unique position that sits somewhere in between research and commercial venture. They are often capable of moving at commercial speed, they develop and test for specific, on-the-ground applications intended for a wide audience, but they don't have the same conflict of interest that arises from a tension between revenue making and human needs. By way of example, the Young and Well Cooperative Research Centre, an $80 million Australian initiative that we mentioned in previous chapters, has goals that bridge applied research and dissemination of the innovation produced. It is jointly funded by the Australian government, nonprofits, for-profit companies, and universities.

As another example in this sector, the Inspire Foundation, a nonprofit organization dedicated to mental health prevention and promotion among young people, designs and develops a variety of technologies for its work. With branches in the United States, Ireland, and Australia, it is funded through donation and sponsorship. Their User Experience teams partner with clinical and research psychologists and psychiatrists and engage in participatory codesign with the young men and women who are their users. Critical as this sector is to moving efforts forward, progress would surely be limited if positive-computing projects could be pursued only through nonprofit- or government-funded means. Enter social business.

Positive Computing as Social Enterprise—the Fourth Sector
The past few years have seen the emergence of an exciting new development within business that has come in response to the need for sustainable self-funding businesses that don't place revenue optimization as their overriding priority. The social enterprise is described by Nobel Peace Prize winner and social business advocate Muhammad Yunus as "a non-dividend company created to solve a social problem. Like an NGO, it has a social mission, but like a business, it generates its own revenues to cover its costs. While investors may recoup their investment, all further profits are reinvested into the same or other social businesses."[2]

At the time of writing, the Social Enterprise Alliance[3] boasts a membership of more than 900 social enterprises, service providers, investors, corporations, public servants, academics, and researchers. According to a recent article in *The Guardian*, "Social enterprises are now part of the fabric of British life. Up and down the country they are tackling social problems and improving communities, people's life chances and the environment, and then reinvesting their profits back into the business or the local community. They are a growing, exciting and vibrant part of the mix in today's jobs market."

Examples include everything from innovating in the recycling industry to preserving local history and culture, providing early childhood education, and solving hard problems in disadvantaged communities. Even large corporations are contributing by helping these enterprises get off the ground and gain momentum. And there is a certification body to monitor the integrity of the social enterprise promise.[4]

A recent *Wikipedia* definition of the term *social enterprise* makes the link to positive computing explicit: "A social enterprise is an organization that applies commercial strategies to maximize improvements in human and environmental wellbeing, rather than maximising profits for external shareholders."[5] This growing sector of industry seems not only too long awaited and incredibly sensible, but also particularly well suited as a funding model for work in positive computing. The crossover model has the potential agility and self-reliance of a business with the applied community-driven approach of a nonprofit combined with a freedom from many of the burdens and conflicts of interest posed both by profit-driven business and by the nonprofit need to continually seek funding. No endeavor is free from interests and biases, but in a decade or two we may

Caveats, Considerations, and the Way Ahead

look back and find that progress in positive computing owes much to the social enterprise.

Benefits of Wellbeing

In addition to the conditions and factors of wellbeing, it's also worthwhile to look at the consequences of it, not only for their own sake, but because they can be important tools for justifying positive-computing initiatives. The list of follow-on benefits can be helpful in getting support for positive computing from those for whom greater happiness is not enough reason to make or fund change.

There are myriad positive follow-on effects to increased wellbeing that appear in other aspects of a person's overall life experience, and although some of these effects are wholly intuitive, others may surprise.

For example, among the more familiar is evidence that happiness seems to boost the immune system and thus reduces incidents of illness (O'Leary, 1990). Inversely, stress increases the chance of illness over both the short and the long term. As such, work in wellbeing can be viewed as a kind of preventative health-care measure in addition to an overall aid to treatment.

More surprising, perhaps, is the finding that developing emotional resilience in school children, in addition to increasing their wellbeing, improves their academic performance (Durlak, Weissberg, Dymnicki, Taylor, & Schellinger, 2011), a consequence that might very well help motivate stakeholders to invest more time and funding into wellbeing efforts in education.

Similarly, many workplaces invest in wellbeing programs based on the understanding that happier workers perform better on the job and engage better with their team. This relates to the research that shows positive emotions increase creativity and open-ended thinking.

It has also been acknowledged that wellbeing promotion prevents mental health crises that contribute to societal problems such as crime and substance abuse. Supporting psychological wellbeing allows people to reach their potentials in all walks of life, and the "side effects" can be impressive.

The Positive-Computing Project

As a researcher/practitioner team, we are keenly aware that both a scientific underpinning *and* a realistic set of professional best practices are necessary

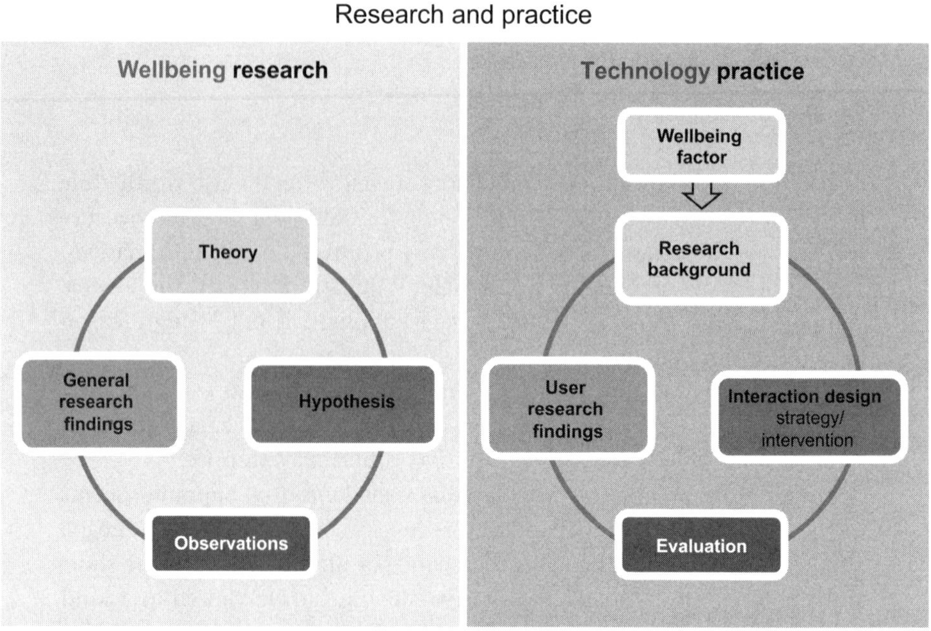

Research and practice

Figure 12.1

for positive-computing projects to thrive. But how do we move these two sides of the equation toward a common goal to produce a sustainable product?

One challenge is the different work processes and time scales that exist between research and practice, and we have presented them in figure 12.1. Whereas scientists are accustomed to the process on the left-hand side, practitioners tend to follow processes like the one on the right.

We have made an effort to highlight the commonalities between the two, but the differences remain substantial. As mentioned earlier, a scientific project is likely to be scoped a number of years in advance. Needless to say, predicting the nature of available technologies that far ahead is terribly difficult, if even possible.

In contrast, those working for a company or nonprofit organization will have project timelines measured in months at most, and the greatest challenge is often having enough time to gain an up-to-date understanding of the area and then measure its actual impact over the longer term. Just measuring impact, if one includes the follow-up evaluations at 3, 6, or 12 months, will often call for a period of time longer than the length of many projects themselves.

Therefore, positive computing will likely require new approaches that allow developers to measure the impact of new technologies more accurately as well as new approaches to funding that allow researchers greater flexibility to adapt to technology change. Perhaps there is a sweet spot to be found through cross-sector collaboration.

Some of the organizations we work with include in their project proposals both traditional user research deliverables such as personas and participatory design findings and positive-computing variables such as initial wellbeing evaluations and a related psychology literature review to support the project design.

Despite the challenges, an increasing number of projects continues to emerge from multiple sectors all the time. We highlight some of these projects together with emerging work in multiple fields at positivecomputing.org.

. . .

One thing unique to humans is our relentless tendency to reshape and redesign our world. We don't leave things well enough alone. We continually optimize. We have used this incredible ingenuity for both dramatically good and not so good results. Interest in positive computing seems to reflect a growing collective consensus that the impact we have as a species has not been as socially clever or insightful as it needs to be to improve things in the long term. Instead, we have sometimes confused money with happiness and "biggering" with bettering (as Dr. Seuss put it in *The Lorax*), and we have neglected the bigger picture, the broader community, and the planet itself (a.k.a. "everything"). As a result, we seem now to be conceding that to make optimum progress for optimum sustained happiness and with minimum disaster and suffering, we'll need to turn to principles of sustainable wellbeing for ourselves, our communities, and our planet. Upon these efforts rests the future of technology.

Buddhists have a core practice called *metta* that involves actively wishing that all beings be well and happy. We asked one Buddhist monk how we could genuinely wish happiness for even our greatest enemies, even the really terrible people, such as mass murderers and torturers. He pointed out that if these people were genuinely well and happy, they would not be terrible. The idea that things such as violence and cruelty, in addition to other types of suffering such as anxiety or depression, are born of illbeing, seems to make a study of wellbeing a clear imperative, a necessary contribution to the solving of all human-derived social problems, and a compliment to so many of the other endeavors in this direction. Perhaps by attending to wellbeing through technology, we are in some small way spreading this intention—that all beings everywhere be well and happy—and in so doing, we are improving (one iteration at a time) the way we as humans change our world.

Notes

1. Find out how many slaves work for you at slaveryfootprint.org.

2. From Muhammad Yunus's social business website (yunussb.com).

3. See www.se-alliance.org.

4. For this certification body, see socialEnterpriseMark.uk.

5. "Social Enterprise," *Wikipedia*, http://en.wikipedia.org/wiki/Social_enterprise, accessed February 10, 2014.

References

Brynjolfsson, E., & McAfee, A. (2011). *Race against the machine*. Lexington, MA: Digital Frontier Press.

Corporates can ensure social enterprises mean business. (2013). *The Guardian*, August 7. Retrieved from http://www.theguardian.com/social-enterprise-network/2013/aug/07/big-businesses-support-social-enterprises.

Davidson, R. J., & Begley, S. (2012). *The emotional life of your brain: How its unique patterns affect the way you think, feel, and live—and how you can change them*. New York: New American Library.

Diener, E., & Diener, M. (2009). Cross-cultural correlates of life satisfaction and self-esteem. In E. Diener (Ed.), *Culture and well-being: Collected works of Ed Diener* (vol. 38, pp. 71–91). The Hague: Springer.

Durlak, J. A., Weissberg, R. P., Dymnicki, A. B., Taylor, R. D., & Schellinger, K. B. (2011). The impact of enhancing students' social and emotional learning: A meta-analysis of school-based universal interventions. *Child Development*, *82*(1), 405–432.

Elsass, P., Carlsson, J., & Husum, K. (2010). Spiritualitet som coping hos tibetanske torturoverlevere (Spirituality as coping in Tibetan torture survivors). *Ugeskrift for Laeger*, *172*(2), 137–140.

Fox, S., & Duggan, M. (2013). *Tracking for health*. Washington, DC: Pew Research Center's Internet & American Life Project.

Friedman, B. (1996). Value-sensitive design. *Interaction*, *3*(6), 16–23.

Glasgow, R. E., & Rosen, G. M. (1978). Behavioral bibliotherapy: A review of self-help behavior therapy manuals. *Psychological Bulletin*, *85*(1), 1–23.

Norman, D. A. (2007). *The design of future things*. Philadelphia: Perseus Books.

O'Leary, A. (1990). Stress, emotion, and human immune function. *Psychological Bulletin*, *108*(3), 363–382.

Rogers, Y., & Marsden, G. (2013). Does he take sugar? Moving beyond the rhetoric of compassion. *Interaction*, *20*(4), 48–57.

Rotman, D. (2013). *How technology is destroying jobs* (pp. 28–35). MIT Technology Review. Cambridge, MA: MIT.

Schimmack, U., Radhakrishnan, P., Oishi, S., Dzokoto, V., & Ahadi, S. (2002). Culture, personality, and subjective well-being: Integrating process models of life satisfaction. *Journal of Personality and Social Psychology*, *82*(4), 582–593.

Sin, N. L., & Lyubomirsky, S. (2009). Enhancing well-being and alleviating depressive symptoms with positive psychology interventions. *Journal of Clinical Psychology*, *65*(5), 467–487.

Index

academic performance, 49
addiction, 118
affect
  detection, 68, 69
  expression, 69
affective awareness, 157–159
affective computing, 51, 68–72, 212, 215
  attentive interfaces and, 69, 70
  wellbeing measurement and, 98
affective sharing, 205
affiliative and soothing system, 116–118
affirmations, 140
affluence, 42
aggression, 146, 147
alcohol education, 51
altruism, 2, 87, 230
  design for, 246–249
  games for developing, 245, 246
  inspiration and, 240–242, 248
  motivations for, 236–238
  research on, 230–236
  strategies for cultivating, 242, 243
  wellbeing and, 240–242
altruistic interaction, 244
altruistic punishment, 233, 234
anger, 235
anxiety, 71, 163
Apple devices, 1, 123
appreciation, 171

architecture, 54, 55
Aristippus, 18
attention, 181
attentive user interfaces, 69, 70
aural feedback, 194, 195
automated caregivers, 221, 222
autonomy, 21, 22, 50, 149, 150, 193, 194, 261–265
awareness, 181

"back-stage" behaviors, 139
Bailenson, Jeremy, 249, 250
bandwagon phenomenon, 266, 267
Beck Depression Inventory, 184
behavior change technology, 72, 73
Bejewelled, 147
Benson, Buster, 164, 165
Bickmore, Timothy W., 221, 222
biofeedback, 189
biology, 28–32
blame, 233, 247
boyd, danah, 58, 59
brain structure, 28, 29
brain waves, 183
Breath–Walk Aware System, 192
Brin, Sergey, 1
Buddhist psychology, 26–28, 179, 180, 207, 230, 234, 235, 277
Burns, Jane, 56, 57
business, 53, 54

Candy Crush, 147
Caruso, David R., 173, 174
Center for Epidemiological Studies–
  Depression Scale, 5, 16, 17, 93–96, 184
character traits, 157
circumstantial conditions, 82
cognitive awareness, 157, 158
cognitive behavioral therapy (CBT), 137, 155, 159, 160–162, 182
cognitive dissonance theory, 139, 140
cognitive distortions, 160, 163, 164
commercial industry, 272, 273
communication
  Internet-based, 99
  nonverbal, 209
  technology-mediated, 209, 217, 223, 224
compassion, 87
  definition of, 229, 230
  design for, 246–249
  detection of, 248
  evolutionary basis for, 237
  in face of injustice, 234–236
  fairness and, 232–234
  inspiration and, 241
  research on, 230–236
  strategies for cultivating, 242, 243
  technologies for development of, 243–246
  uniqueness of, 231–233
  wellbeing and, 238–240
compassion-focused therapy, 170, 232, 242, 243
competence, 21, 22, 50, 263–265
computer-based CBT, 160–162
contentment, 112
contingent rewards, 133
control issues, 265, 266
control-value theory, 50
Cook, Tim, 1
cooperation, 56
cooperative games, 127

culture, 268, 269
customer service, 218, 219
cyberbullying, 91, 208
cyberpsychology, 24
cybertherapy, 71
Csikszentmihalyi, Mihaly, 223, 224

Dalai Lama, 26, 27, 234–237
data
  automated collection of, 103
  log, 65, 66
  utilization of, 75, 76
Data-Rich Writing Studios, 70
Declaration of Independence, 46
denial, 114
depression, 30, 113, 163
  measurement of, 94–96
  therapy for, 137
descriptive investigation, 99
deservedness, 247
design, 54, 55
  active, 91, 217, 218
  affiliative, 117
  for compassion and altruism, 246–249
  dedicated, 91
  for empathy development, 217–220
  for engagement and flow, 149–151
  integration of, into practice, 89–91
  measures of effective, 99–102
  for mindfulness, 193–196
  motivating, 149
  for positive emotions, 121–128
  preventative, 90, 91, 208, 209, 217, 218, 243
  for reflection, 172
  for self-awareness, 172
  technology, 6, 7
  visceral, 124
Design for Good, 1
determinant factors, 83–89
Diagnostic and Statistical Manual, 16

digital streets, 164
digital technology, hedonic wellbeing and, 18, 19
Diplomacy, 148
disruption, 271
distractions, 150, 194
diversity-training software, 249–250
dog-collar technology, 66
dopamine, 116
driverless cars, 31, 32, 261, 262
drives, 135

Easterlin Paradox, 42, 43
Echo, 163, 164
economics, 41–44, 52
education, 47–52, 82, 188, 189, 213
elevation, 240–242
embodied experiences, 191, 192, 219, 246
emoticons, 217
emotional computers, 69
emotional intelligence, 2, 25, 26, 158, 159, 173, 216
emotional labor, 114
emotions, 68, 69
emotive behavioral therapy, 160
empathetic play, 219, 220
empathic joy, 207, 208
empathy, 87, 114, 203–224
  affective, 205
  in animals, 203, 204
  art and, 206, 207
  capacity for, 204
  cognitive, 205
  vs. compassion, 231, 232
  components of, 205
  decline in, 208, 209
  designing for, 217–220
  developing, 205–209
  group, 215–217
  measurement of, 209, 210
  parental, 206
  strategies for fostering, 210–212

technologies for development of, 212–217
training, 233
understanding, 204, 205, 215
Empathy Quotient, 209, 210
employee wellbeing, 53, 54
employment, 43
empowerment, 263–265
enablers, 24
endorphins, 116
engagement, 87, 131
  agentic, 144
  behavioral, 143, 144
  cognitive, 143
  definition of, 143
  designing for, 149–151
  emotional, 143
  flow and, 144, 145
  games and, 145–149
  measurement of, 145
  modes, 144
entries, 122, 123
environment, 82, 83
environmental psychology, 54, 55
epigenetics, 30
esteem, 135
ethical dilemmas, 270
ethnographic studies, 102, 103
eudaimonia, 111
eudaimonic psychology, 21–23, 31
Eurobarometer, 21
excitement and drive system, 116–118
exits, 122, 123
experience sampling, 98, 155
experiential awareness, 157–160
experimental designs, 99–102
exposure therapy, 71
external regulation, 133
eye gaze, 252

fairness, 232–234
family, 82
fear, 235

Fitbit, 65, 66
Five Facets Mindfulness Questionnaire, 190, 191
5-HTT promoter, 30
fixed mindset, 136, 137
flourishing, 14, 23–25, 36, 112, 113
flow, 131, 144, 145, 149–151
Fogg model, 141
framework for positive computing, 81–89
Freiburg Mindfulness Inventory, 184
Frontiers, 214
"front-stage" behaviors, 139
funding issues, 271–275

Gallup-Healthways Well-Being Index, 45
game addiction, 147, 148
Games for Change, 1
Games+Learning+Society, 193
gamification, 134, 149
genetics, 29, 30
gesture devices, 125
Global Assessment of Functioning Scale, 5, 6, 16, 17
Glosser, 70
goal setting, 137–139
goal-setting tools, 137
Google, 1
graphics, 218, 219
gratitude, 87, 171
greenwashing, 266, 267
gross national happiness, 2
growth mindset, 136, 137
guided sessions, for mindfulness training, 190
guidelines for positive computing, 92, 93

HAPIfork, 192
happiness, 2, 3, 14, 18, 19, 111
  measurement of, 5, 44, 45, 96–104
  success and, 112
  wealth and, 42, 43

Happy Planet Index, 21, 44
haptic feedback, 194, 195
Harper, Richard, 4
Hay Day, 244
hedonic psychology, 17–23, 30, 31, 110, 111
hedonic treadmill concept, 20
human complexity, 259, 260
human–computer interaction (HCI), 1–5
  methods, 98–104
humanistic values, 2
human-machine legacy, 3–5
human potential, 2
human replacement, 261–263
human-resource systems, 54
humility, 88
Huppert, Felicia, 35, 36

identification, 133
ill-being, measurement of, 94–96
imagery, 243
Immordino-Yang, Mary-Helen, 251, 252
impression management, 139
income comparison, 43
inspiration, 230, 240–242, 248–252
Inspire Foundation, 33, 34
instrumental software, 103
integration, 133, 134
interaction design for children, 51, 52
interactionism, 150, 151
interactionist theory, 139
interest, 112
interjection, 133
International Classification of Diseases, 16
Interpersonal Reactivity Index, 209
interventions
  for cultivating altruism and compassion, 242, 243
  for fostering mindfulness, 187–189
  Internet-based, 6, 92, 93, 160–164

positive psychology, 119, 120
preventative, 93, 94
promotional, 93, 94
introspection, 156
ironic processes, 67, 140, 259

joy, 112
judgment, 247

Kahneman, Daniel, 18, 19, 42, 181
Kindle, 122
kindness, 117
Krusche, Adele, 197, 198

Lake Woebegone Effect, 169, 170
leadership styles, 216, 217
learning technologies, 50–52
LEGO therapy, 211
libertarian paternalism, 47, 142
life logging, 65
life-long learning, 82
life satisfaction, 19, 21, 96, 97
love, 112, 135

Mayer-Salovey-Caruso Emotional Intelligence Test, 25
McGonigal, Jane, 127, 128
media, 223
meditation, 187, 188, 233, 242
memory, 115, 116, 121–123, 126
mental capital, 54
mental flexibility, 205
mental health
  promotion, 47–50, 93, 94, 163, 164
  technology for, 71, 72
mental illness, 29, 30
  prevention of, 47, 93, 94
  in youth, 47
metacognition, 157, 158
middle path, 21
Mindful Attention Awareness Scale, 184, 185

mindfulness, 18, 49, 87, 138, 156, 159, 179–198
  concept of, 179, 180
  design for, 193–196
  digital technology for, 189–193
  impact of, on wellbeing, 184–187
  measurement of, 183, 184
  as natural attribute, 182
  neuroscience of, 182, 183
  as nonjudgmental attention, 181, 182
  psychology of, 181, 182
  scientific evidence for, 180
  strategies for fostering, 187–189
  training, 187, 188
Mindfulness-Based Cognitive Therapy, 180, 182
Mindfulness-Based Stress Reduction program, 27, 180
Mindfulnets, 190, 191
Mind Reading, 211, 213
mind wandering, 185, 186
minimalism, 193, 194
mobile devices, 1
mobile phones, for therapy interventions, 72
Mood-Meter, 98, 103
moral agency, 73
motivation, 21, 22, 118, 131–141, 265, 266
  designing for, 149
  extrinsic, 132–134
  intrinsic, 132–135
  social, 139, 140
multitasking, 186, 187

narcissism, 167–169, 208
narrative, 218, 219
national progress measures, 44, 45
National Well-Being Programme, 44
needs
  hierarchy of, 135, 136
  physiological, 135

negative emotions, 29, 111–114
negative mood, 19
neural networks, 28, 29
neuroscience, 28–30
New Economics Foundation, 44
Nicholas, Jonathan, 33, 34
nonjudgment, 181, 182, 195, 196, 235, 247
nonstriving, 193, 195
Norman, Don, 125, 126
nudge theory, 72, 142

obedience, 206
observation methods, 99
online disinhibition effect, 91
organizational psychology, 53, 54
outcomes, 24
oversensitivity, 218

Page, Larry, 1
pain avoidance, 132
Panic Attacks, 162
paternalism, 265, 266
Peace-Maker, 214
peak–end rule, 115, 116, 122
people-replacement model, 161
Perceived Stress Scale, 190, 191
PERMA model, 23–25, 81
personal informatics, 64–68, 165–167
personality, 29, 30, 82
persuasion, 141, 142
persuasive technology, 72, 73
pity, 231
pleasure principle, 132
plurality, 224
Positive and Negative Affect Scale, 184
positive computing
  definition, 2
  framework, 81–89
  guidelines, 92, 93
  projects, 275–277

positive emotions, 21, 28, 29, 49, 50, 109–128
  authenticity and, 113–115
  broadening and building effect of, 111–113
  designing for, 121–128
  groups of, 116–119
  increase in wellbeing and, 110–115
  strategies for cultivating, 119–123
  tipping point of, 113
positive mood, 19
positive psychology, 2, 23–25, 83, 100, 101
positive technology, 24, 25
positive thinking, 113–115
positivity, 138
power, 265, 266
pride, 112
privacy, 260, 261
productivity, 262
prosociality, 146, 208, 244, 245
psychiatrists, 16
psychiatry, 5
psychoeducation, 242, 243
psychological wealth, 112
psychology, 5, 6
psychotherapy, 71
public goods, 273, 274
public policies, 44–47, 52

quality-of-life instruments, 5, 6
quantified self, 65–67, 155, 165–167, 259

ReachOut.com, 33
reality, 115, 116
reflection, 70, 71, 156, 157
  concept of, 159
  design for, 172
  vs. direct instruction, 164, 165
  on self, 167, 168
  strategies for, 159, 160

technology-mediated, 163, 164
  for wellness and wellbeing, 165–167
reframing, 114
regulatory processes, 205
relatedness, 21, 22
relationships, 82
remembered experiences, 121–123, 126
repression, 114
resilience, 2, 30, 87, 117
Rogers, Yvonne, 75–76
role-playing games, 214, 219
Roots of Empathy, 211, 212
rumination, 18, 167, 168

safety, 135
Savannah, 104
science, 271, 272
security, 260, 261
self-actualization, 136
self-awareness, 87, 155–160, 168, 169, 172, 205
self-compassion, 169–171, 196, 230, 239, 240
self-criticism, 167, 168
self-determination theory, 21, 22, 31, 32, 50, 72
self-esteem, 169–171
self-focus, 167
self-help books, 23, 24
self-help exercises, 100, 101
self-knowledge, 64, 65, 156, 157
self-reports, 115
self-tracking, 66–68, 165–167
Seligman, Martin, 23
serotonin, 30
sleep cycles, 29
Smiling Mind, 190
social-emotional learning, 49, 188, 189
social enterprises, 1, 274, 275
social exchange theory, 237

social good, 1, 2
social influence, 134, 140
social knowledge, 73, 74
social media, 58, 59
  altruism in, 244
  data, wellbeing measurement and, 43, 44
  empathy and, 208
  enterprise-level, 53
  motivations for using, 134
  plurality and, 224
  wellbeing and, 52
social networks, 120, 190, 191
social phenomenology, 139
social responsibility, 88
social science, 52, 53
Sonic Cradle, 191, 192
spirituality, 87
strengths, 24
stress reduction, 72, 190, 238
SuperBetter, 113
sympathy, 232, 237
synchronous movement, 248, 249

talents, 24
technological progress, 3–6
television, 53
Tenacity, 193
test anxiety, 50
theory of planned behavior, 72
Tibetan refugees, 235, 236
transparency, 266
transtheoretical model, 72
trolling, 91
Turing, Alan, 4
tutoring systems, 51

ubiquitous computing, 63, 64, 258
Uplifted, 149
user experience, 5, 99, 121, 124, 150, 217
user interfaces, 70, 143, 243

utility
  instant, 97
  objective, 97, 98
  remembered, 97, 98
  total, 123
UX for Good, 1

values-sensitive design, 32, 73–75
videogames, 120, 121, 127, 128
  for empathy development, 213–215
  for mindfulness, 192, 193
  prosocial, 146, 208, 244, 245
  violent, 146, 147
  wellbeing and, 146–149
virtual reality, 71, 215, 243, 245, 246
volunteering, 238–241

wealth, happiness and, 42
web 2.0 applications, 99
wellbeing, 13–40
  balance and, 269, 270
  benefits of, 275
  components of, 17, 21, 22, 29, 36
  conditions and factors that influence, 82, 83
  design for, 31, 32
  ecology of, 270
  eudaimonic, 21–23, 30, 31
  as experience of positive emotion, 17–19, 22, 23, 30, 31
  as flourishing, 23–25
  government policy and, 44–47
  happiness and, 43
  hedonism and, 110, 111
  measurement of, 2, 5, 6, 17–21, 35, 36, 93, 96–104
  medical model of, 15–17
  at molecular level, 30, 31
  multidisciplinary approach to, 41–59
  paradigms of, 14, 15
  as physiologically identifiable, 28–30
  place and, 54, 55
  promotion of, 16
  questions about, 13–16
  subjective, 19–21, 35, 36, 96, 97, 158, 159
  technology and, 2–5
  in technology research, 63–76
  in the workplace, 53, 54
Wiener, Norbert, 4
Williams, J. Mark G., 197, 198
wisdom, 88
Wisdom 2.0, 1
work-related stress, 54, 190
World Happiness Report, 21, 43, 45, 97
writing, 70, 71

Young and Well Cooperative Research Centre, 15, 56

Zombies, Run!, 131, 149